Environmental Management

Revision Guide for the IEMA Associate Membership Exam and NEBOSH Diploma in Environmental Management

Written by Adrian Belcham, author of the *Manual of Environmental Management*, this is the essential guide to prepare you for the IEMA Associate Membership Exam and NEBOSH Diploma in Environmental Management.

Through the inclusion of revision tips, exam guidance and self-test questions, this guide will consolidate your understanding of environmental management and help you to prepare for your exam. It provides full coverage for both the IEMA and NEBOSH exams and includes coloured sections to help you to identify the content relevant for each qualification.

- Small, handy size, ideal for when on the move.
- Includes illustrations and tables to improve understanding.
- Written by an expert tutor of IEMA and NEBOSH environmental qualifications.

Adrian Belcham is a Chartered Environmentalist, full member of the Institute of Environmental Management and Assessment, and an EARA accredited Environmental Auditor. Since 1998, he has been lead tutor of the UK-based training organisation Cambio Environmental Ltd, which specialises in the delivery of IOSH, IEMA and NEBOSH environmental qualifications.

Environmental Management

Revision Guide for the IEMA Associate Membership Exam and NEBOSH Diploma in Environmental Management

Adrian Belcham

First published 2015
by Routledge
2 Park Square, Milton Park, Abingdon, Oxon OX14 4RN

and by Routledge
711 Third Avenue, New York, NY 10017

Routledge is an imprint of the Taylor & Francis Group, an informa business

British Library Cataloguing-in-Publication Data
A catalogue record for this book is available from the British Library

Library of Congress Cataloging-in-Publication Data
[CIP data]

ISBN: 978-1-138-77534-3
ISBN: 978-1-315-77384-1 (ebk)

Typeset in Univers LT by
Servis Filmsetting Ltd, Stockport, Cheshire

Printed and bound in Great Britain by
TJ International Ltd, Padstow, Cornwall

Contents

CHAPTER 1

Using this guide

This guide has been prepared to assist students in the preparation for two examinations with considerable overlap in subject matter – the IEMA Associate Membership exam and the NEBOSH Diploma in Environmental Management. It is intended to be used in conjunction with the syllabi for each programme (produced by the IEMA and NEBOSH) as well as the *Manual of Environmental Management* by Adrian Belcham. This guide provides condensed coverage of each syllabus intended to serve as a revision synopsis with full explanations provided in the Manual.

The guide begins with a section on the IEMA and NEBOSH examinations and strategies for preparing for each. The qualifications are set at different academic levels and it is important for students to be clear about the expected standards for each.

The main body of the guide is divided into 10 chapters that correspond to the chapters of the *Manual of Environmental Management* and also to the IEMA syllabus elements. Each of these 10 chapters contains 4 sections:

1. Concise notes, tables, images and revision task boxes are provided to aid students with revision. The NEBOSH syllabus covers a wider spectrum of issues than the IEMA one and the green boxed sections are relevant to NEBOSH students only (see pages 52–4, 71–4 and 100–5).
2. IEMA self-test questions and syllabus links.

3. NEBOSH self-test questions and syllabus links.
4. Further reading suggestions.

Appendix 1 and 2 contain a series of specimen answers to exam questions based on those set by the IEMA and NEBOSH. These are linked to the self-test questions and provide students with a clear benchmark of the standards to which they should be aligning themselves.

Appendix 3 is a glossary of environmental terms which provides an alphabetical quick reference to many of the key concepts covered within the guide.

Appendix 4 is a revision tool in the form of a legal summary table provided for you to complete as part of your exam preparation process. This has proven to be a very effective learning tool and 'memory jogging' reference over the years. You will need to refer to Chapter 5 of this guide plus the *Manual of Environmental Management* or alternative references to complete the table with your own key words.

CHAPTER 2

The IEMA and NEBOSH examination requirements

This chapter is organised into the following sections:

2.1 The differences between the IEMA and NEBOSH examination levels

It is important to recognise that although this guide and the accompanying Handbook are intended for use by both IEMA Associate and NEBOSH Diploma students, the two qualifications have different syllabi, examination formats and assessment standards.
It is important to be clear about the revision standards that you are trying to achieve. Passing the IEMA Associate membership exam requires less scope and depth of knowledge from students than the NEBOSH Diploma. Green shaded text boxes in Chapters 3–12 of this guide highlight those areas of information required only by NEBOSH Diploma students. It should be assumed that all other information in the guide and in the accompanying Manual is relevant to both IEMA and NEBOSH students. NEBOSH Diploma students need to recognise that, compared with IEMA students, they will be required to:

- consider more information (albeit across similar topic areas)
- understand and recall all topic areas to a greater degree of detail
- cultivate the ability to incorporate multiple areas of the syllabus within the context of a given scenario
- write more extensively, coherently and with a higher degree of interpretation in the examination.

2.2 Common issues – command words and revision tips

Examination command words

The terms and definitions in Table 2.1 are the key command words used in both the IEMA and NEBOSH examinations. Learn to recognise them and pay close attention to them when preparing your answer.

Revision tips

- Little and often is the golden rule for revision – most people find it more effective to spend an hour a day rather than one full day a week.

Table 2.1 Examination command words (based on NEBOSH guidance)

<table>
<tr><th>Command word</th><th colspan="2">Meaning</th></tr>
<tr><td>Identify</td><td colspan="2">Normally a name, word or phrase will be sufficient, provided the reference is clear. For example: identify the regulator responsible for noise and statutory nuisance.</td></tr>
<tr><td>Explain what is meant by</td><td colspan="2">Give a recognised definition of something or describe it in such a way so that it is distinguished from anything else. For example: explain what is meant by sustainable development.</td></tr>
<tr><td>Outline</td><td colspan="2">State and make clear, or give the most important features of something. Outline is equivalent to a thin description, but involves more than simply listing (more detail than define but often expanded bullet points will be appropriate).</td></tr>
<tr><td>Describe</td><td colspan="2">Give enough detail about an item for a reader to have a clear picture of it. Describe normally requires a short paragraph and involves more writing than outline.</td></tr>
<tr><td>Explain</td><td colspan="2">A detailed answer proving clarity of interpretation. Often associated with 'how' or 'why' – includes giving reasons for something, linking cause and effect, drawing parallels, pointing to relationships or showing how theory can be applied. Explain normally involves giving more detail than 'describe' and will often include words and phrases like 'because', 'since', 'so that', 'in order to'. . . etc.</td></tr>
<tr><td>Assess</td><td colspan="2">Subject something to critical analysis in order to make a judgement about its value, use, suitability, integrity or accuracy.</td></tr>
<tr><td>Interpret</td><td colspan="2">Interpret a set of data by describing the main trends, highlighting any anomalies, then providing an explanation of the data based on knowledge and understanding of the particular subject area.</td></tr>
<tr><td>Propose</td><td>Suggest actions</td><td rowspan="2">These terms are often used in context based questions.</td></tr>
<tr><td>Apply</td><td>Suggest how to use</td></tr>
</table>

- Don't be too dependent on one information source – if when looking at the relevant syllabus you find areas that you feel are insufficiently covered in this study guide or the Manual, or that don't give you exactly what you want, e.g. a diagram or summary – use the references provided and read around the subject. At the very least it is a good idea to familiarise yourself with the Environment Agency and DEFRA websites.

- Remember that both the IEMA and NEBOSH examinations are designed to test your understanding – you cannot expect to adequately prepare for the exam by learning by rote. Use past paper questions and examiners reports where available as a way to gauge your understanding.
- **Write some long hand answers** rather than just bullet point plans. Initially do this with reference to your notes but then try some 'blind' and finally 'timed, blind' answers. Many candidates that fail the exam do so because of inadequate interpretation and adherence to the questions. Read the question carefully, write a quick answer plan (especially for the NEBOSH exam) and then refer to both question and plan when writing your long hand answer – **practise this!**
- Prepare a **revision plan** (see Table 2.2) dividing up the relevant syllabus into the number of weeks' revision time you have allocated yourself. Remember to give yourself reasonable goals, breaks and review points using timed past paper questions. Mix up your approach – reading, preparing summary notes or mind maps, speaking presentations explaining a topic area, past paper questions as answer plans or full written answers etc.

While relying too heavily on 'question spotting' in exams is a recipe for disaster, many examinations work from a limited question bank and there is inevitably some repetition. IEMA and NEBOSH are no exception in this respect and we would be unwise not to recognise that fact. In this study guide you will find examples of questions that appear regularly. It is a good idea to incorporate consideration of your own answers and any marking schemes or model answers provided as part of your exam preparation.

2.3 IEMA examination tips

Key features of both the IEMA online and paper-based examinations include:

- 2.5 hours long, i.e. 15 minutes per question – no pre-start reading time.
- The exam format is 'open book' but is expressly designed not to

Table 2.2 A sample revision timetable

	Mon	Tues	Wed	Thurs	Fri	Sat	Sun
Week 1	Global and local envl issues	Law – permitting	Sustainability	Envl assessment techniques	off	Monitoring and measurement	Envl assessment techniques
Week 2	Pollution control techniques	Law – water & contaminated land	Sample exam questions	Law – air and climate	off	Envl assessment techniques	Sample exam questions
Week 3	EMS and KPI's	Law – waste	Sample exam questions	Carbon mgt and footprinting	off	Law – development control & nuisance	Law – civil cases & timed sample Q's
Week 4	Law – haz subs & general powers	Procurement & Biodiversity	Sample exam questions	Review of revision summaries	off	Revision summaries & blind exam questions	Revision summaries & blind exam Q's

be a 'look up' exercise. However, you will of course be able to look up legal references, and use glossary definitions, checklists etc. that you know to be relevant to the question.

- For online examinations, answers are typewritten, while paper-based examinations typically require handwritten submissions.
- 10 questions each worth 12 marks.
- All questions must be answered – no choice available.
- Questions are asked in relation to English legislation but can be answered in relation to your local legislation where relevant.
- Reference should be made to relevant legislation and workplace examples wherever appropriate.
- The 10 questions relate to each of the 10 learning outcomes in the IEMA Associate syllabus (and to chapter headings in the manual), i.e. you will receive one question in relation to each learning outcome.
- The qualification is pitched at the National Qualifications Framework level 4 standard which may be considered roughly equivalent to an 'A' level assessment.

The IEMA Associate syllabus is available from www.iema.net. It is recommended that you become familiar with it. Table 2.3 illustrates the links between the learning outcomes specified in the 2013 syllabus and the chapters of this guide and the *Manual of Environmental Management*.

Setting the scene for the IEMA online exam

Plan for your exam to reduce the likelihood of interruptions, distractions etc. Consider the following:

- **Familiarise yourself with the online format** – by watching the IEMA exam tutorial at www.iema.net and by logging in but not starting the exam once you have received your registration details.
- **Location** – home or work – you'll need to be somewhere where you can be undisturbed for 2.5 hours, be able to spread out your reference materials and where you are not likely to be disturbed by excessive noise etc.
- **Timing** – choose a time that means you are less likely to be disturbed – taking an afternoon off work to do the exam at home

Table 2.3 The IEMA syllabus and links to this guide (based on IEMA Associate Syllabus 2013)

IEMA Associate learning outcome	Revision guide chapter	Manual chapter
1. Understand environmental and sustainability principles	3	1
2. Understand environmental policy issues	4	2
3. Understand key environmental legislation and compliance measures	5	3
4. Understand environmental management and sustainable development in a business context	6	4
5. Be able to collect, analyse and report on environmental information and data	7	5
6. Be able to apply environmental management and assessment tools	8	6
7. Be able to analyse problems and opportunities to deliver sustainable solutions	9	7
8. Be able to develop and implement programmes to deliver environmental performance improvement	10	8
9. Be able to communicate effectively with internal and external stakeholders	11	9
10. Be able to influence behaviour and implement change to improve sustainability	12	10

while no one else is in the house, for example. But equally choose a time that you know you're likely to be at your best – some people prefer early morning and some late evening etc. Also consider internet reliability – if this is an issue for you choose a time of day when the connection is most reliable.

- **Preparation** – view the exam overview video so that you know what to expect. Set up your log-in in advance so that you know everything works before you go to do the exam.
- **Organise your references** – you won't be able to spend much time looking things up so make 'key facts' accessible so that you know in advance what you're likely to look up and exactly where it is.

During the exam:

- Stay calm and read each question slowly and carefully!
- If struggling to understand what the question refers to remember the link to the syllabus learning outcomes – this may point you in the right direction. However, don't be too limited in your thinking – it may well be the case that information from other syllabus categories is relevant.
- Be disciplined in what you look up – you do not have time and, in most cases anyway, will not find the answers in your books.
- Use the command words and mark allocations to guide the amount of detail you give for each question.
- Keep focused on the time allowance – 15 minutes per question – don't over-run.
- If for any reason the exam gets interrupted – note the time, cause and duration of the interruption. If it is possible to do so, continue and complete the exam but then report the interruption to IEMA as soon as you are done.

2.4 NEBOSH examination tips

Key features of the NEBOSH diploma examination include:

- 3 hour paper based examination with a 10-minute pre-start reading period.
- Candidates answer 5 questions from 8 provided, i.e. 36 minutes per question.
- The exam format is 'closed book' (i.e. no reference materials are allowed) and require handwritten answers.
- Questions are often context based and cover more than one area of the syllabus – this requires candidates to apply knowledge and demonstrate understanding rather than simply recall information.
- Reference should be made to relevant legislation, management standards and workplace examples wherever appropriate.
- The qualification is pitched at the National Qualifications Framework level 6 standard which may be considered roughly equivalent to a Bachelor's degree level assessment.

Table 2.4 The NEBOSH syllabus and links to this guide (based on the NEBOSH Diploma in Environmental Management syllabus 2010)

Revision guide chapter	Manual chapter	NEBOSH Diploma syllabus element references	
3	1	Element 1 (1.1, 1.2, 1.3 [in part])	
5	3	Element 5 (5.1–5.4) Element 6 (6.1–6.2) Element 7 (7.1) Element 8 (8.1–8.2) Element 9 (9.1–9.2)	Element 10 (10.1–10.2) Element 11 (11.1–11.3) Element 12 (12.2) Element 13 (13.1–13.2) Element 14 (14.1–14.2)
6	4	Element 1 (1.3 [in part])	
8	6	Element 2 (2.1–2.2) Element 3 (3.2)	Element 4 (4.2)
9	7	Element 3 (3.1–3.3) Element 4 (4.1–4.2) Element 9 (9.3–9.5) Element 10 (10.3–10.4)	Element 11 (11.4) Element 12 (12.1–12.2) Element 14 (14.2) Element 15 (15.1–15.2)
11	9	Element 7 (7.2–7.3)	

The NEBOSH Diploma syllabus is available from www.nebosh.org.uk. It is recommended that you become familiar with it. Table 2.4 illustrates the links between the learning outcomes specified in the 2013 syllabus and the chapters of this guide and the *Manual of Environmental Management*.

Before the exam:

- Give yourself plenty of time to get to the examination centre – if appropriate travel to the venue the night before, to give yourself as relaxed a start to the exam morning as you can.
- Don't try to do too much in the way of revision or reading on the morning of the exam – key notes are OK as a reminder but to think that you can take on more information this late in the day is probably deluding yourself. You're better off going into the exam with a clear, calm frame of mind rather than trying to 'cram' right up to the last minute.

During the exam:

- NEBOSH allow a 10-minute 'reading period' at the start of the exam. You are not allowed to write anything during this period but this is when you should read through the whole paper and select the 5 questions you will attempt out of the 8 on offer. Don't panic if your 5 favourite topics don't come up – you are likely to get a mix of questions which may be thought of in categories:
  - Class A – definitely good questions for you – do one of these first to get you flowing but don't be tempted to spend more than the allotted time on these 'favourites'
  - Class B – second choice questions
  - Class C – 'not unless I really, really have to' questions

It's OK to have a mix of classes that you will answer. Generally it's a good idea to leave Class C questions until last – not because you should spend least time on them but because you don't want a difficult question to put you off your Class A and B questions. Writing your 1 to 5 order down as soon as you're allowed to write can save you revisiting your choice between each question and thereby save valuable time.

- Once you are allowed to start writing read the whole question **again** carefully (many delegates underline key words) and make sure you answer in the separate sub-sections given (if appropriate) rather than merging the answers together.
- Identify the command words which provide a guide to the depth of answer required.
- Spend a few moments planning your answer – a basic essay plan may simply be a series of bullet points or headings but writing them down first can help ensure a logical structure that is focussed on answering the question. It is easy otherwise to get side-tracked, diverted or go into automatic 'brain dump' once you start writing.
- Organise your answer in a way that makes distinct points clearly visible, e.g. paragraphs, bullet points, headings, clear use of section numbers provided in the question etc. Remember your examiner is comparing your answer to a

marking scheme – the easier you can make it for them to cross reference your answer with the marking scheme the more marks you're likely to get.

- Make sure your answer is legible! If an examiner cannot read your answer easily you are unlikely to win marks.
- When you start writing the full answer keep an eye on three things:
  - The clock – stick to your allotted time per question (36 minutes maximum)
  - Your answer plan
  - The question!

What the NEBOSH examiners say

The following paragraphs are from a NEBOSH examiner's report:

> In order to meet the pass standard for this assessment, acquisition of knowledge and understanding across the syllabus are prerequisites. However, candidates need to demonstrate their knowledge and understanding in answering the questions set. Referral of candidates in this unit is invariably because they are unable to write full, well-informed answers to the questions set.

Key point – it doesn't matter if you know your stuff really well if you cannot interpret a question and apply your knowledge to the specific context given!

NEBOSH has also stated that the key factors that they consider necessary to exam success may be weighted as follows:

> Innate ability (20%); Good knowledge of the subject (30%); Strong exam technique (50%).

Key point – like it or not we have to make learning and practising exam technique a key component in our preparation for the NEBOSH assessment.

It is important to recognise that the answers to the NEBOSH questions will not simply be a 'look-up' in your training course notes but may involve selecting and interpreting information from a variety of locations and perhaps using your own experience and judgement to provide a comprehensive answer. This point is emphasised in another paragraph from an examiner's report:

> Some candidates find it difficult to relate their learning to the questions and as a result offer responses reliant on recalled knowledge and conjecture and fail to demonstrate any degree of understanding. Candidates should prepare themselves for this vocational examination by ensuring their understanding, not rote-learning pre-prepared answers.

The marking process

The process for setting and marking NEBSOH examinations is lengthy and involves considerable efforts to standardise the approach taken by different examiners. The basis of the approach is a marking scheme that is prepared collectively by the examination team and must then be applied by each examiner. Understanding how this process works is very important in maximising your chance of winning marks.

Key point – the examiner marking your paper is likely to spend around 5 minutes per question (and possibly less). Knowing this means we have to plan to make it as easy as possible for them to award us marks against their marking scheme guidance!

CHAPTER 3

Environmental and sustainability principles

This chapter is organised into the following sections:

3.1 Revision notes

3.1.1 Concepts and definitions

For a comprehensive glossary of terms see Appendix 1 in the *Manual of Environmental Management*.

Environment definitions:

1. Everything, including human beings – the whole of the planet acting as a linked and interdependent whole.
2. 'Surroundings in which an organisation operates including air, water, land, natural resources, flora, fauna, humans and their inter-relationships.'

Key components – Physical environment, Biological environment, Human environment.

Ecosystem definition – 'communities of interdependent organisms and the physical environment that they inhabit'.

Ecosystem examples – woodland, lakes, forests, moorland etc.

Biodiversity – the variety of life in a given area typically expressed in relation to ecosystems, species and genetic variation within species.

Environmental impacts include pollution impacts and resource depletion impacts.

Resource depletion occurs when we consume non-renewable resources (e.g. fossil fuels) or potentially renewable resources (e.g. fish stocks and woodlands) that are inappropriately managed.

Pollution impacts require the presence of a ***source – pathway – receptor*** link.

Pollution impacts may be acute or chronic and can occur at a variety of scales – local, regional and global. Linkages between different elements of the environment may give rise to secondary and/or cumulative impacts.

Aspect definition – Those elements of an organisation's activities, products and services which can interact with the environment.

- Hazardous material usage
- Air emissions
- Effluent discharges

- Solid waste
- Raw material usage
- Energy usage
- Land usage
- Noise/nuisance.

Classic aspects checklist – mnemonic – SHARLENE or REALHENS

Impact definition – environmental changes occurring as a result of an organisation's aspects. Impacts affect receptors and as with aspects there is a generic list of environmental receptors that might be affected by an organisation's activities:

- Human beings
- Flora and fauna
- Soil
- Water
- Air and climate
- Landscape
- Cultural heritage and material assets.

Task

Write a sentence or short paragraph to provide examples of the following terms:

A pollution impact

___

___

___

A secondary impact

___

___

___

An indirect impact

____

____

____

3.1.2 Natural cycles

Natural cycles describe the flow of materials or energy through the biological and physical environment. The linkages created by these cycles often give rise to knock on effects (secondary impacts) from an initial impact in one part of the environment. Figures 3.1–3.5 show the key natural cycles affected by human activity.

3.1.3 Environmental impacts – an overview

Air pollutants

Table 3.1 Air pollutants – key sources and impacts (author)

Pollutant	Key sources	Key impacts
Oxides of Nitrogen (NOx)	Combustion of fossil fuels	Acid rain, photochemical smog, climate change
Sulphur Dioxide (SO2)	Combustion of fossil fuels	Acid rain
Carbon Monoxide (CO)	Combustion of fossil fuels (especially vehicle emissions)	Human health impacts
Ozone (O3)	Secondary pollutant arising from photochemical smog	Human health impacts
Particulate Matter	Agricultural and industrial dust plus diesel emissions from vehicles	Nuisance and human health impacts
VOCs	Solvent usage	Climate change, ozone depletion, photochemical smog and human health impacts
Carbon dioxide	Combustion of fossil fuels	Climate change

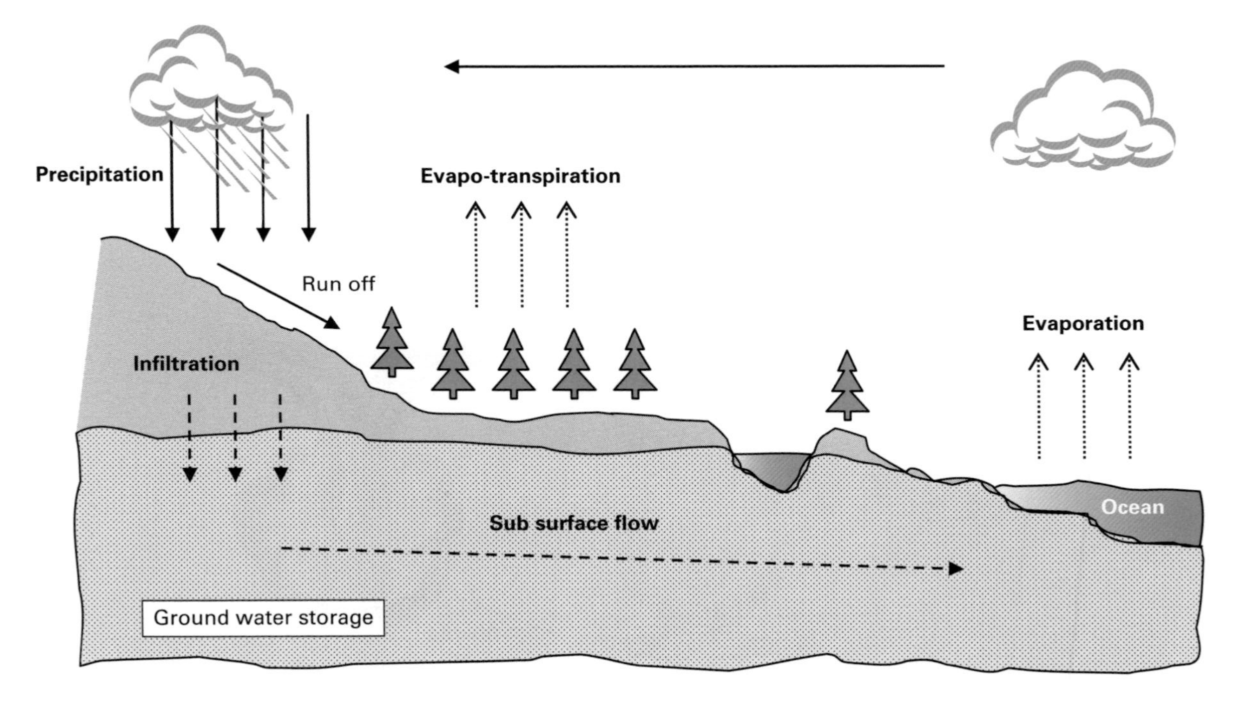

Figure 3.1 The hydrological cycle

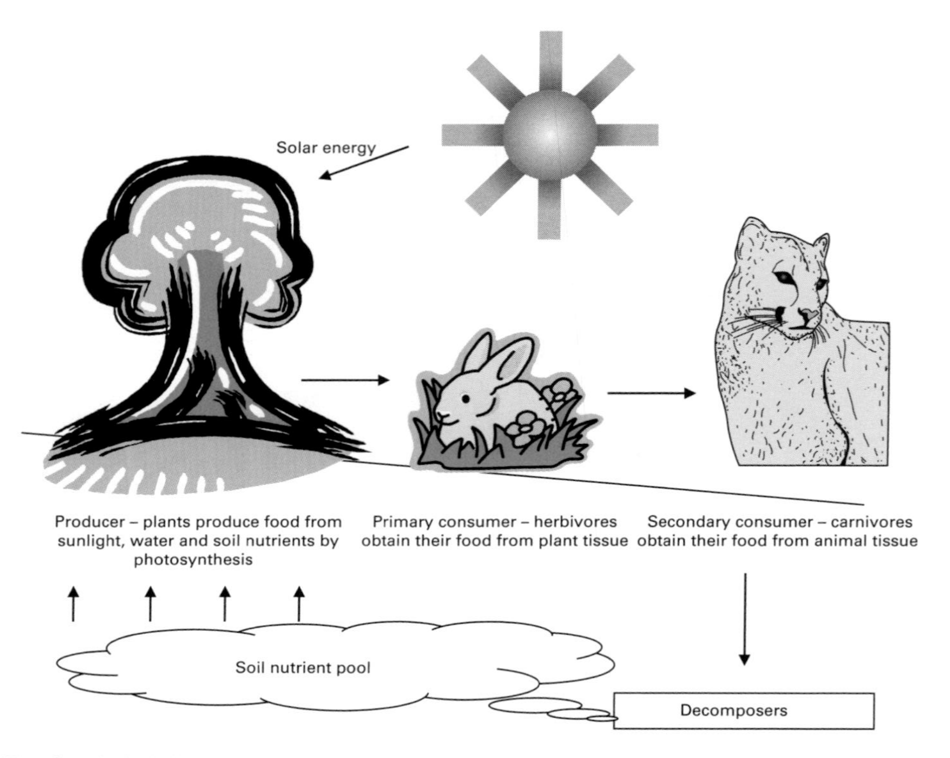

Figure 3.2 The food chain/energy cycle

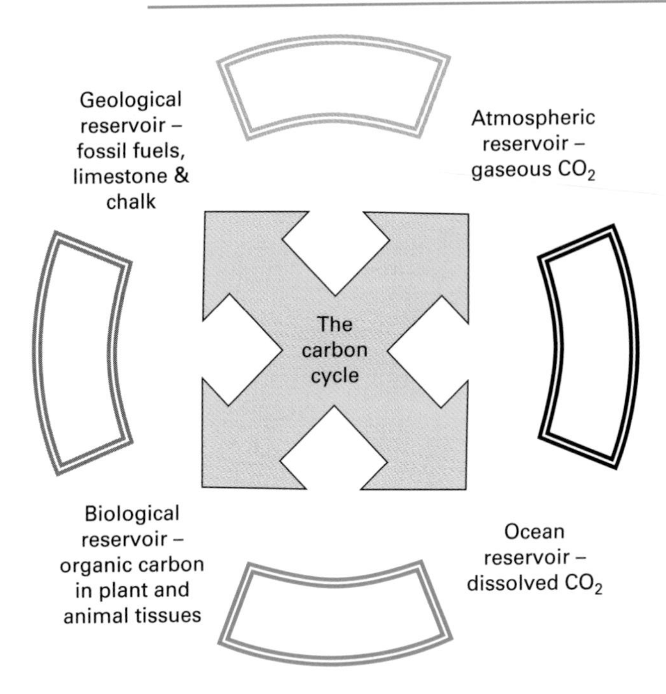

Figure 3.3 The carbon cycle

Climate change

Cause – Exaggeration of the Greenhouse Effect

Key pollutants – Greenhouse gases including carbon dioxide, methane, and nitrous oxide

Key sources – Combustion of fossil fuels

Key impacts – Extreme weather events, sea level rise, coastal flooding, changes in vegetation zones and hydrological systems etc.

Ozone depletion

Cause – Chlorine and bromine reactions with ozone in upper atmosphere

Key pollutants – Examples: CFCs, HCFCs, halons, methyl bromide, carbon tetrachloride

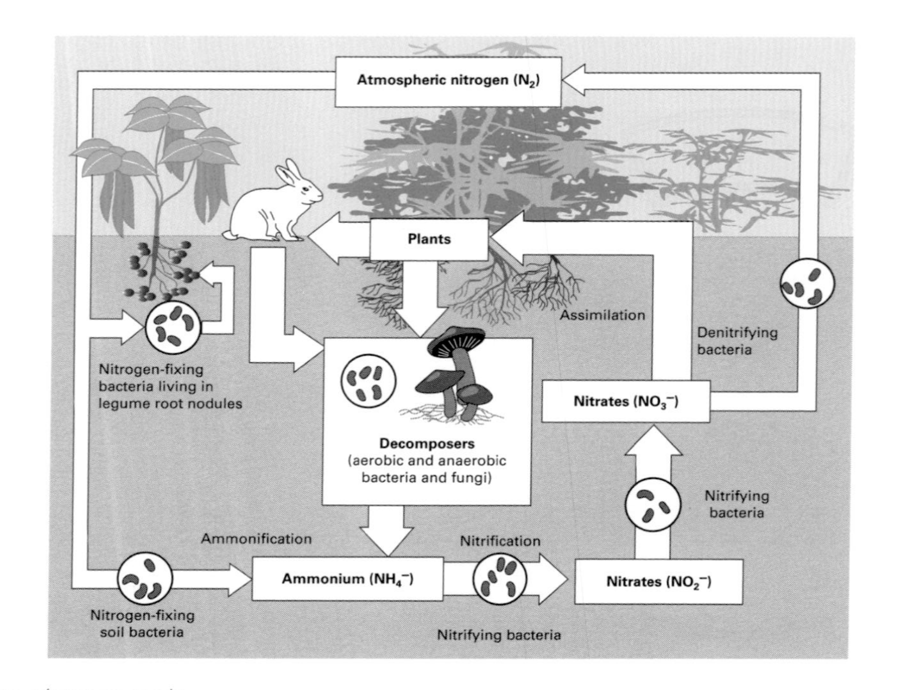

Figure 3.4 The nitrogen cycle

Source: US Environmental Protection Agency

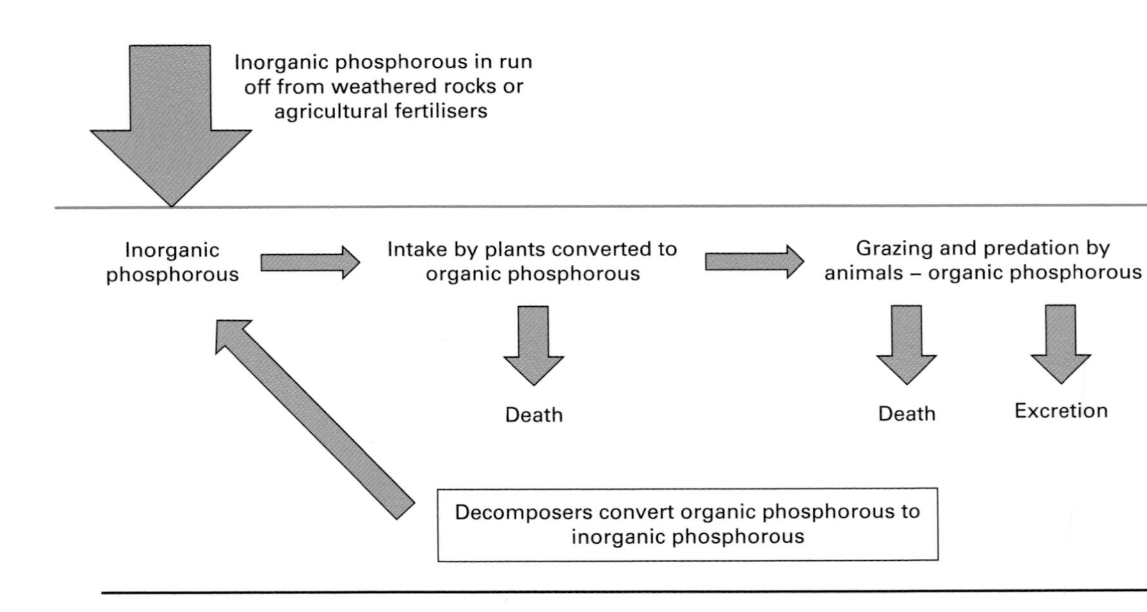

Figure 3.5 The aquatic phosphorous cycle

Characteristics of ozone depleting substances:

- Persistent – only broken down by exposure to strong u/v light
- Contain chlorine or bromine atoms
- Free Cl and Br atoms create chain reactions leading to ozone depletion

Key sources – Refrigerants, fire suppressants

Key impacts – Increase in ground level exposure to ultraviolet radiation and associated human, plant and animal health impacts

Acid rain

Cause – Acid gases combining with water in atmosphere to produce acidic precipitation

Key pollutants – SO_2 and NOx

Key sources – Power stations and transport

Key impacts – pH change in soil and water with knock on damage to flora, fauna, agriculture and infrastructure

Photochemical smog

Cause – Chemical reaction of air pollutants under appropriate climatic conditions

Key pollutants – Oxides of nitrogen and VOCs

Key sources – Vehicle emissions

Key impacts – Production of toxic secondary pollutants including ozone, aldehydes, ketones and peroxyacetyl nitrites (PANs)

Water pollutants

Table 3.2 Water pollutants – key sources and impacts (author)

Pollutant group	Effect
Nitrates and phosphates: run-off from intensive agriculture using nitrate based fertiliser	Can promote eutrophication leading to excessive algae growth which reduces light to plants below the surface. Dissolved oxygen levels can be reduced through bacteria living on the dead plant.

Organic wastes: including untreated sewage, food industry wastes etc.	Reduces dissolved oxygen levels which causes changes in aquatic flora and fauna. Sewage contains bacteria and other pathogens harmful to humans.
pH: industrial discharges	Excessive acidity or alkalinity can be toxic to fish.
Oil: accidental seepage from storage tanks, or deliberate dumping	Poisons aquatic life. Prevents oxygen absorption so the dissolved oxygen levels will decline and may inhibit flora and fauna.
Suspended solids: coal washing, china clay	Increased turbidity reduces light levels. Smothering of river beds.
Thermal: industrial cooling water	Elevated water temperature reduces dissolved oxygen levels while increasing the rate of chemical and biological activity.
Toxic compounds: industrial	Toxicity to aquatic and other life depending on levels. Toxic materials can be stored up in fish and concentrated.
Endocrine disruptors: oestrogen and oestrogen mimicking substances such as phthalates	Disturbance to reproductive capacity of individual organisms or whole species. Mutagenic effects. Human health impacts.

Groundwater pollution

Cause – Spills and leaks of pollutants from a wide variety of sources

Key impacts – Loss of valuable supplies of potable water. Transfer of contaminants over a wide area and possibly reaching surface water. Very difficult and expensive to clean up

Freshwater availability

The ability to meet increasing demands for high quality fresh water is threatened by over abstraction and pollution of surface and ground water supplies.

This has significant implications for both human communities and sensitive ecosystems.

Management of natural resources

Energy

Cause – Reliance on fossil fuel energy supply

Key impacts – Pollution associated with fossil fuel energy. Increasing pressure on availability of resources. Lack of development of renewable alternatives such as wind, solar, wave or hydroelectric schemes

Contaminated land

Many current and historic sources of land contamination in the UK. Typical sites associated with contaminated land include waste, chemicals and fuel storage/usage locations.

Availability of productive land

Cause – Population pressure, urbanisation, climate change and poor land management

Key impacts – Reduced 'food security' for an increasing population. Pollution and biodiversity impacts associated with current methods to increase productivity

Biodiversity loss

Cause – Unprecedented species loss associated with poor land management. Biodiversity is the variety of life in a given area, described in terms of habitats, species and genetic diversity within species

Key impacts – Loss of organisms critical in maintaining ecological systems. Loss of species with direct human use applications, e.g. medicines from plants

Chemicals exposure

Cause – Of the 100,000 or so chemicals thought to be in production today – very little or no information is available on 85 per cent of them

Key impacts – The human threats posed by individual chemicals or complex mixtures range from sensitisation and associated allergic responses, through infertility to cancers. The truth of the matter is that for many chemicals and chemical mixtures we simply do not understand the risk. Particular concern about **Persistent Organic Pollutants** (POPs) – because of their ability to bioaccumulate; and **endocrine disruptors** because of their intergenerational impacts

Waste

Cause – Large quantities of waste generated annually. Over-reliance on landfill disposal

Key impacts – Landfill impacts include:

- Leachate
- Contaminated land
- Landfill gas
- Land use pressure.

Human community issues

A range of issues affecting both small numbers of individuals and human society in general. Broadly grouped under the following headings:

- Nuisance – noise, dust, odour – infringement of use of property, loss of sleep etc.
- Infrastructure / amenity impacts – influx of construction personnel – pressure on accommodation etc.
- Cultural heritage impacts – physical destruction, pollution, visual proximity and social influx / dominance examples

3.1.4 Sustainability – the place of humans in the environment

Sustainability is a description of environmental balance (including human society's place in that balance) – sometimes the term 'environmental sustainability' is used to denote this meaning.

Sustainable development is the process of transforming human society into a form that allows 'sustainability' to exist. It involves consideration of social and economic issues (as well as the environmental ones) that will need to be addressed along the way.

An ***ecosystem*** may be defined as:

> A community of interdependent organisms and the physical and chemical environment that they inhabit.

Ecosystem services may thus be defined as:

> Those attributes of ecosystems that enable or facilitate human life and well-being.

Provisioning services:

- food (including seafood and game), crops, wild foods and spices
- water
- minerals (including diatomite)
- pharmaceuticals, bio-chemicals and industrial products
- energy (hydropower, biomass fuels).

Regulating services:

- carbon sequestration and climate regulation
- waste decomposition and detoxification
- purification of water and air
- crop pollination
- pest and disease control.

Supporting services:

- nutrient dispersal and cycling
- seed dispersal
- primary production.

Cultural services:

- cultural, intellectual and spiritual inspiration
- recreational experiences (including ecotourism)
- scientific discovery.

International commitments to sustainable development under the UN Commission for Sustainable Development and the Earth Summit series. Agenda 21 was an agreement reached by heads of state at the first Earth summit in Rio de Janeiro in 1992 – it is also known as the blueprint for sustainable development.

Significant debate is emerging about the paradoxical goals of economic growth and sustainability. While sustainable development wisdom suggests a three pronged approach – environment, society and economy – it is suggested that a new model of economic health (other than growth) may be required to give sustainability a chance of success.

Task

Write a paragraph explaining why you think the international community has agreed a vision of sustainable development that encompasses change in relation to social, economic and environmental factors.

3.2 IEMA self-test questions

Question 1

a) Describe either the carbon cycle or the nitrogen cycle. [4 marks]
b) With reference to two contrasting examples, describe how human interventions impact upon natural cycles. [8 marks]

Question 2

a) Describe the main causes of climate change. [6]
b) Identify 3 potential impacts of climate change and describe the consequences. [6]

Question 3

a) Describe the main causes of acid precipitation. [6]
b) Identify 3 potential impacts of acid precipitation and describe the consequences. [6]

3.3 NEBOSH self-test questions

Question 1

a) **Explain** what is meant by the term '*biodiversity*'. [6]
b) **Explain** why the maintenance of appropriate levels of biodiversity within ecosystems is considered important from a human perspective. [14]

Question 2

Describe the ways an organisation might attempt to minimise its local and global adverse impacts on plant and animal communities (biodiversity). [20]

3.4 Further information

Topic area	Further information sources	Web links (if relevant)
Aspects and impacts	ISO14001:2004 Environmental management systems – requirements with guidance for use	www.iso.org
Sustainability	United Nations sustainable development knowledge platform	sustainabledevelopment.un.org/
	The Sigma Project	www.projectsigma.co.uk/
	The Natural Step	www.naturalstep.org/
	Forum for the Future	www.forumforthefuture.org/
Sustainability indicators	DEFRA publication – UK sustainable development indicators 2013	www.gov.uk/defra
	Welsh Assembly publication – Sustainable development indicators 2013	wales.gov.uk

CHAPTER 4

Environmental policy

This chapter is organised into the following sections:

4.1 Revision notes

4.1.1 Generic principles

The Oxford English dictionary defines a policy as:

> a course or principle of action adopted or proposed by an organization or individual.

Policy generally drives the creation of legislation and fiscal measures such as tax regimes and incentives such as grants and education programmes. Arising from Agenda 21 (the internationally agreed sustainable development strategy) we have a number of fundamental principles that increasingly shape and inform international, national and corporate policy.

- **Polluter pays principle** – polluter pays for abatement, clean-up and regulatory costs
- **Preventative approach** – avoid rather than mitigate environmental harm
- **Precautionary principle** – minimise risk and act where suspicion of harm exists even if absolute proof unavailable
- **Best available techniques** – agree and require best practice standards by all potential polluters
- **Producer responsibility** – relates to waste materials and requires those involved in the production of key products to be accountable for their end of life recovery, recycling and safe disposal
- **Life cycle thinking and best practicable environmental option** – demands consideration of the impacts of design changes and operational improvements to ensure that in overall terms there is a net benefit to the environment

4.1.2 Policy instruments – an overview

Policies of themselves do not create change. They require mechanisms of implementation that bring them to the attention of appropriate parties and cultivate desirable behaviour while discouraging undesirable behaviour. Once such positive and negative perceptions are adopted as a widely held social norm, change can be said to have occurred.

Legal instruments – widely used but require well-resourced regulation

Fiscal measures – taxes, tax relief and grants plus charging for regulatory time

Market based measures – examples include the EU emissions trading scheme, corporate reporting requirements etc. that create financial and public relations incentives for organisations to behave in a particular way

Education and awareness schemes – typically targeted at particular sectors and often particular issues, these schemes are often essential supporting mechanisms for the other categories of measures. UK examples include industry focused programmes run by the Carbon Trust, WRAP etc.

Task

Prepare an overview of the key policy instruments relevant to your company under the four headings:

- Legal instruments – select the highest profile/priority examples only
- Fiscal measures
- Market based measures
- Education and awareness schemes.

Table 4.1 Advantages and disadvantages of non-legal policy instruments (author)

Advantages of non-legal instruments	Disadvantages of non-legal instruments
• Allow individual flexibility and choice of action based on personal circumstances • Taxation can raise revenues to be invested in causes supporting the environmental issue in question • Has a greater tendency toward 'self-regulation'	• May lead to organisations simply 'paying to pollute' or ignoring best practice guidance • If trading prices, arrangements or tax levels are set too low there may be little incentive to change behaviour • If trading prices, arrangements or tax levels are set too high there may be

Advantages of non-legal instruments	Disadvantages of non-legal instruments
• More likely to result in active engagement in solution finding/ performance improvement (the more we do the more we save)	trade disadvantages created compared with operations in jurisdictions with no similar mechanisms • Can be cumbersome to administer with costs associated with accounting and assurance that reduce the benefits of any revenues raised

4.1.3 Policy instruments – examples

Table 4.2 Examples of environmental policies, principles and implementation measures (author)

Environmental issue	Example of related policy	Key principles incorporated	Examples of implementation instruments
Climate change	The UK Low Carbon Plan 2011	Polluter pays principle, precautionary principle	The Climate Change Levy The Green Deal The carbon reduction commitment (CRC) scheme Mandatory greenhouse gas reporting for plcs
Ozone depletion	The Montreal Protocol on substances that deplete the ozone layer	Producer responsibility	EC Regulation 744/2010 on ozone depleting substances
Trans boundary air pollution	The convention on long range trans-boundary air pollution	Polluter pays, producer responsibility, BPEO	EU Integrated Pollution Prevention and Control regime, 1999
Water pollution	The International convention for the prevention of pollution from ships (Marpol convention)	Polluter pays principle and producer responsibility	The UK 'Merchant Shipping (Prevention of Pollution by Sewage and Garbage from Ships) Regulations 2008'

Environmental issue	Example of related policy	Key principles incorporated	Examples of implementation instruments
Biodiversity loss	The Convention on Biological Diversity	The precautionary approach	National enforcement of the Convention on International Trade in Endangered Species UK local and regional biodiversity action plans
Pollution associated with hazardous waste disposal	The strategy for hazardous waste management in England 2010	Polluter pays, preventative approach, life cycle analysis and BPEO	The Hazardous Waste (England & Wales) Regulations 2005

4.2 IEMA self-test questions

Question 1

With reference to a named waste policy, describe how the polluter pays principle is implemented. [12]

Question 2

a) Identify four advantages and four disadvantages of using fiscal instruments as an environmental policy instrument. [4]
b) Describe two fiscal instruments currently in use that deal with different environmental problems, stating the country in which they apply. [8]

Question 3

Describe the main policy instruments used in environmental policy implementation. [12]

4.3 NEBOSH self-test questions

Question 1

a) **Summarise** what is meant by the term *'best available technique'* in the context of the Pollution Prevention and Control Act 1999. [5]

b) **Describe** the scope and key requirements that may be imposed by a permit to operate a part A1 'regulated activity' as listed in the Environmental Permitting (England & Wales) Regulations, 2010. [15]

Question 2

a) **Outline** the ways in which a national supermarket chain may contribute to causing the phenomenon known as the 'Greenhouse Effect'. [10]
b) **Explain** the main elements of the UK Climate Change Policy (the Low Carbon Plan) and describe examples of policy instruments employed in its support. [10]

4.4 Further information

Topic area	Further information sources	Web links (if relevant)
Generic principles	Agenda 21	sustainabledevelopment.un.org/
Specific policy examples (national)	UK government policy web portal	www.gov.uk/government/policies
Specific policy examples (international)	UNEP Global Outlook on Sustainable Production and Consumption policies	www.unep.org/resourceefficiency/

CHAPTER 5

Environmental law

This chapter is organised into the following sections:

5.1 Revision notes

5.1.1 An overview

Common/civil law – involve 'judicial precedents', i.e. decisions made in deciding disputes between two parties.

Legislation/stature – law enacted by parliament following consultation and review. Most environmental stature is enactment of European Directives and Regulations.

Table 5.1 Key regulators in England (author)

Legislative Body	Areas of Responsibility
Department for Environment, Food and Rural Affairs (DEFRA)	Drafting and implementing legislation Negotiating national agreements Formulation of policy Provision of guidance to other regulatory bodies Provision of scientific advice
Department of Energy and Climate Change	The UK Climate change programme
Environment Agency	Environmental Permitting Water Waste Producer responsibility Radioactive substances Contaminated land (special sites)
Local Authorities	Environmental Permitting Clean air Air quality Nuisance Contaminated land
Sewerage Undertakers	Effluent to sewer
Natural England	Conservation and land management issues

Penalties for non-compliance

Prosecution, statutory notices (including civil sanctions), clean-up costs, compensation claims, reputation damage.

The Environmental Civil Sanctions (England) Regulations 2010

Provide an alternative to prosecution to secure binding and effective action to prevent or remediate pollution. **For a specified list of offences**, regulators are able to use alternative sanctions with legitimate businesses who are trying to do the right thing. Six types of sanction: Compliance Notice, Restoration Notice, Fixed Monetary Penalty, Variable Monetary Penalty, Enforcement Undertaking, Stop Notice.

The Environmental Damage (Prevention and Remediation) Regulations, 2009

Regulators may raise a 'remediation notice' to force clean up or pollution prevention as follows:

Environment Agency	In relation to permitted activities and/ or water damage/biodiversity damage in waters other than the sea
Natural England	Damage to biodiversity
Local Authorities	Land damage
Secretary of State	Damage at sea

5.1.2 Environmental permitting

EU IPPC Directive implemented via the Pollution Prevention and Control Act 1999 and the Environmental Permitting (England & Wales) Regulations, 2010.

Permit required (unless exempted by the regulations) to:

(a) operate a regulated facility; or
(b) cause or knowingly permit a water discharge activity or groundwater activity.

Regulated facilities classified as:

Part A1 processes – require a permit from the EA and are regulated in relation to emissions to air, land and water as well as with regard to other operational conditions such as noise, energy efficiency and waste minimisation.

Part A2 processes – require a permit from the Local Authority. They may be regulated against the same range of operational conditions as A1 processes and, if discharges to water are involved, then in England and Wales the EA must be consulted.

Part B processes – require a permit from the Local Authority but are only regulated in relation to their discharges to air.

Permits may be standard, bespoke or consolidated. Permit conditions may include:

- emission limit values to land, air and water (air only for part B processes)
- resource efficiency conditions (relating to raw materials, wastes and energy)
- protection measures for soil and groundwater
- site reinstatement requirements
- nuisance control limitations
- monitoring and operational management requirements (including incident prevention and response).

Operators must prove they meet Best Available Techniques (BAT) in order to receive a permit. In the event of non-compliance regulators may raise enforcement, prohibition or revocation notices.

For A1 processes, the Environment Agency uses the Environmental Protection Operator and Pollution Risk Appraisal (EP OPRA) methodology to set permit fees based on perceived pollution risk.

Task

Prepare a short presentation to a senior management team on the Environmental Permitting process using the following headings to structure your talk:

- UK and European legal context
- Who requires an environmental permit?
- What do we have to do to get a permit?
- What permit conditions might we be expected to work within?

5.1.3 Air pollution

Table 5.2 International air pollution control agreements (author)

Montreal protocol	Phase out of ozone depleting substances. Enacted via ***EC Regulation 744/2010,*** supported in the UK by the ***Environmental Protection (Controls on Ozone-Depleting Substances) Regulations 2011*** Cessation of new product, cessation of recycled product, ban on usage.
Convention on long range trans boundary air pollution	Originating from concerns about acid rain this agreement aims to limit the spread of pollution across borders – has led in Europe to ambient air quality emission limits plus standardised restrictions on emissions from large combustion processes (via the ***EC framework directive on ambient air quality assessment and management***).
UN framework convention on climate change	This is the international agreement to cooperate to deal with the threat of climate change. Produced the Kyoto protocol (1997–2012) and subsequently the Copenhagen Accord and the plan to generate a binding international system of GHG emissions reductions targets by 2015 aimed at capping global warming. Supported regionally, e.g. the EU Climate Change policy and nationally, e.g. the UK Climate Change Act (2008), and the ongoing low carbon plan.

UK legislation

In addition to controls within the Environmental Permitting regime the following key legislation applies:

Table 5.3 UK air pollution control legislation (author)

Clean Air Act (1993)	Control of dark and black smoke (particulates) from domestic and industrial chimneys. Local authorities can issue an abatement notice where emissions exceed the Ringlemann shade set for the area.
Environment Act (1995) Part IV	Establishes the UK ambient air quality strategy (in line with the European Framework Directive referred to above). Priority pollutants (all traffic related) are monitored by local authorities who must take action where limit levels are breached. Limit levels are set by EU.
Climate Change Act (2008)	Sets out legally binding national GHG emission reduction targets for 2020 (26–32%) and 2050 (80%). Introduced a system of 5 year carbon budgeting and established the semi-autonomous Committee on Climate Change. Targets addressed via the 2011 Low Carbon Plan and initiatives such as the Climate Change Levy, the EU emissions trading programme, the CRC scheme, the Green Deal etc.

5.1.4 Water pollution

In addition to controls on permitted activities outlined within the Environmental Permitting regime the following key legislation applies:

Table 5.4 UK water pollution control legislation (author)

EC Water Framework Directive (2000/60/EC)	Commits European Union member states to achieve 'good' qualitative and quantitative status of all water bodies (including marine waters up to one nautical mile from shore) by 2015. It sets out criteria for assessing water quality and requires states to take a 'river basin' (also known as catchment management) approach to surface water quality management.
Salmon and Freshwater Fisheries Act (1975)	Makes it an offence to discharge any substance to such an extent as to cause the receiving waters to be 'poisonous or injurious to fish, or the spawning grounds, spawn or food of fish'.
Environmental Permitting (England & Wales) Regulations (2010)	All discharges of polluting matter to control waters must be under a permit issued by the Environment Agency unless exempt. Accidental discharges may be prosecuted in this way. Clean rain water run-off exempt.

Water Resources Act (1991)	Defines controlled waters and sets out an Environment Agency licensing regime for abstraction from surface or ground water.
Anti-Pollution Works Regulations (1999)	Empowers the Environment Agency to issue a statutory notice that requires the person served to carry out specified works or actions to deal with pollution of controlled waters. These powers have been enhanced by the Environmental Damage Regulations (2009).
Water Industry Act (1991)	All discharges to the public sewer require prior consent from the local sewage undertaker/water company. Applies to domestic style and trade effluent discharges. If trade effluent contains prescribed substances then water company must seek approval from the Environment Agency prior to issuing consent.

Additional information

The **EC Dangerous Substance Directive (1976)** established the lists of priority water pollutants known as the Black and Grey Lists. The basic requirement is that regulators in all the legislation mentioned above cannot issue a permit or allow operations unless they are confident that strict controls are in place to ensure that levels of these substances in controlled waters are below set standards.

Discharge consent parameters vary widely but might include:

- Quantity – absolute and flow rate
- Dissolved Oxygen (DO)
- Biochemical Oxygen Demand (BOD)
- Chemical Oxygen Demand (COD)
- Total Solids (TS)
- Suspended Solids (SS)
- pH
- Nitrogen
- Temperature
- Metals content
- Oil and grease
- Prescribed substances.

5.1.5 Waste

Waste problems – resource loss/consumption; disposal related impacts (landfill gas and leachate; incineration emissions etc.)

Waste hierarchy – reduce, reuse, recover/recycle, dispose

Table 5.5 International and European areas of legislation (author)

	Key requirements/summary
1989 Convention on the Control of Trans-boundary Movements of Hazardous Wastes and their Disposal (The Basle Convention)	Demands high levels of pre-notification, waste description and assurance of safe transport and disposal arrangements by producers of hazardous waste prior to export.
The 1991 European Framework Directive on Waste (Directive 91/156/EEC revised in 2008)	Defines 'waste' and 'holders' of waste and set out the principles for the duty of care to ensure waste minimisation and appropriate control. The 2008 Directive also sets recycling and reuse targets for specified waste streams.
The 1999 Landfill Directive – Directive 99/31/EC	Relates to the management of landfill sites, including the exclusion of certain wastes to landfill and total reduction targets.
The 1991 Hazardous Waste Directive – Directive 91/689/EEC	Sets out definitions of and requirements for the segregation of hazardous from non-hazardous wastes.
The 1994 Directive on Packaging and Packaging waste – Directive 94/62/EC and updated in 2004	Sets out requirements for minimisation and end of life recovery of packaging waste as well as the elimination of specified hazardous materials from packaging.
The 2000 End of Life Vehicles Directive – Directive 2000/53/EC)	Sets out requirements for the safe disposal of end of life vehicles and establishes recovery and reuse targets. Also specifies the elimination of listed hazardous materials.
The 2003 Waste Electrical and Electronic Equipment (WEEE) Directive and the RoHS Directive (2002/96/EC & 2002/95/EC)	Sets out requirements for manufacturers to recover WEEE and to restrict the use of certain hazardous substances in the manufacture of electrical and electronic equipment.
The 2006 Batteries Directive (Directive 2006/66/EC)	Requires the collection and recycling of all batteries sold within the EU.

Table 5.6 UK waste legislation (author)

	Key requirements/summary
Environmental Protection Act (1990) Part II section 34	Sets out the waste management 'duty of care' for 'holders' of waste, namely: 1. contain it securely to avoid escape; 2. transfer it only to someone with authority to take it (e.g. registered carrier/licensed waste manager), or be allowed to actually take the waste if receiving it; 3. keep appropriate records using transfer notes (including description and information on any potential problems), making sure the documentation is properly completed.
Waste (England and Wales) Regulations (2011)	Require organisations to confirm that they have applied the waste management hierarchy. Introduced a two-tier system for waste carrier and broker registration.
Environmental Permitting (England & Wales) Regulations (2010)	All waste management operations in England and Wales have been required to have a permit or a valid exemption. In some cases eg transfer stations, composting operations and many recycling activities, facilities are eligible for a 'standard permit'. Waste management operations with greater pollution potential, e.g. landfill sites and incineration plants are, however, required to apply for a 'bespoke permit'.
Hazardous Waste Regulations (2005)	Any premises producing more than 500kg of hazardous waste per year must be registered with the Environment Agency as a hazardous waste producer. All hazardous waste must be identified and segregated form other waste types.
Producer Responsibility Obligations (Packaging Waste) Regulations (2010)	Obligated business – sends out more than 50 tonnes of packaging to customers; turnover exceeds £2million. Obligated business must register with the EA, calculate annual packaging recovery obligation and obtain sufficient packaging recovery notes (PRN's) to meet their obligation. Can defer obligation to a compliance scheme but still required to compile accurate accounts.
End of Life Vehicles Regulations (2003)	Requires vehicle producers to set up systems for the free collection of scrap cars (and obtain certificates of destruction). Manufacturers must also design new cars with recycling and reuse in mind, and vehicles must contain an increasing quantity of recycled material.

	Key requirements/summary
Waste electrical and electronic equipment (WEEE) Regulations (2013)	Retailers are responsible for taking back electrical goods which have reached the end of their working life, while producers are responsible for ensuring products are subject to recovery/recycling. Responsibility may be sold on with non-household WEEE – so that business users may have responsibility for recycling at end of life (NB all WEEE is now hazardous waste).
Waste Batteries and Accumulators Regulations (2009)	Producers (or importers) are required to register with a compliance scheme, keep records of the total batteries put into the market and, if more than one tonne of batteries are sold, then the producer pays for recovery and recycling according to market share. The compliance scheme actually administers the recovery process and the Environment Agency regulates the producer registrations, and audits/enforces the whole scheme. Retailers selling 32kg or more of household batteries are required to take back batteries in-store, free of charge, when they become waste.

In addition to controls on permitted activities outlined within the Environmental Permitting regime Table 5.6 summarises the key UK waste legislation.

5.1.6 Contaminated land

No relevant international or European legislation. The problem in the UK is compounded by a long industrial heritage leaving significant areas of 'historic' contamination. Key aim of regime is to assign responsibility for clean-up.

Key legislation – *Contaminated Land (England) Regulations 2006*

Requirements:

Local authorities are required to inspect their areas from 'time to time' to identify contaminated land.

- Remediation Notices are to be issued where appropriate (taking into account costs, seriousness of the problem, and intended use of the site).

- Notices are to be served on 'appropriate persons', i.e. those who caused or knowingly permitted the contamination (class A) or, where this person is not known, the current occupier owner (class B).
  - Remediation may take the form of investigation and monitoring as well as clean-up.
  - Special sites are administered by the Environment Agency, e.g. permitted sites, sites where water pollution is occurring, MOD sites etc.
- Statutory guidance related to the regime was updated in 2012 in an attempt to simplify and increase transparency of contamination classification and remediation requirements.

Additional information

In practice the main impact of this regime has not been through the issuing of remediation notices but on the raising of awareness of liability for contaminated land. Sites are more often subject to remediation as part of planning or permitting processes which provide less exposure to the regulators. Due diligence auditing and assessment of sites has become much more common as a result of this legislation.

Task

Your managing director is considering the purchase of an old industrial site in order to build a new warehouse for your company. The site is unoccupied and the last owner was a land developer who owned the site for two years but never did anything with it. The developer went bankrupt and the site is being sold by the lawyers acting as administrators. There is some suspicion that the site has been used in the past for paint manufacture and more recently for vehicle maintenance and repair. Advise your managing director of the potential legal liabilities that may exist as the new owner of this site – in particular in relation to development and contaminated land.

5.1.7 Nuisance

Often dealt with in civil cases, there is also statute relating to the control of nuisance. In addition to controls on permitted activities outlined within the Environmental Permitting regime, Table 5.7 summarises the key UK nuisance control legislation.

Table 5.7 UK nuisance control legislation (author)

	Key requirements/summary
Environmental Protection Act (1990) Part III	Defines the statutory nuisance powers of local authorities, who can issue abatement notices. Where a local authority is satisfied that the noise, dust, odour etc. from any premises is prejudicial to health or a nuisance, it must serve an abatement notice on the person responsible. This notice may require the abatement of the nuisance or prohibit or restrict its occurrence or recurrence, and may also require the execution of such works and the taking of such steps as are necessary for this purpose.
Noise and statutory nuisance Act (1993) & the Noise Act (1996)	Extend the local authority powers granted under EPA'90 part III to facilitate dealing with street based noise and the confiscation of noise making equipment (in domestic disputes).
Control of Pollution Act as amended (1989)	Construction sites may apply for a section 61 consent issued by the LA, which specifies conditions relating to the control of noisy operations (e.g. timing of works, notification of residents etc.). Failure to meet the terms of the consent, or for noisy construction operations where no consent exists, may lead to the issuing of a section 60 notice which will specify required noise controls.
Town and Country Planning Act (1990)	Consideration of noise within the planning process allows mitigations measures to be required and/or limits to be set for both day-time and night-time exposure to noise as part of the planning consent.

5.1.8 Development control

Two main areas – planning control and protection of wildlife and cultural heritage.

Table 5.8 International conservation agreements (author)

	Key requirements/summary
1992 Convention on Biological Diversity	Three main objectives: 1. The conservation of biological diversity 2. The sustainable use of the components of biological diversity 3. The fair and equitable sharing of the benefits arising out of the utilisation of genetic resources In 2010 the Aichi conference of the parties gave rise to a framework action plan and a series of targets for the 2011–2020 decade.
Convention on the International Trade of Endangered Species (1975)	Places tight controls on the international trade of specimens of selected species. Over the years the list has been extended.
The convention on Wetlands of International Importance – the Ramsar convention (1971)	Has been influential in the development of regional and national legislation to protect areas of wetland of significant importance either because of resident species or because of their value to migratory species.

European legislation

The Habitats Directives (1992 and 1997) plus the Birds Directives (1979 and 2009) have been highly influential throughout Europe in establishing strong conservation status for the most important areas regionally. The special areas of conservation (Habitats Directive) and special protection areas (Birds Directive) essentially 'presume against development'.

UK legislation

Table 5.9 UK planning and conservation legislation (author)

	Key requirements/summary
Town and Country Planning Act (1990) as amended	Requires planning permission for all but exempt developments and demands that planning authorities (and statutory consultees) consider applications in such a way as to minimise nuisance and environmental impact.

	Key requirements/summary
Town and Country Planning (Environmental Impact Assessment) Regulations (2011)	Require completion of an 'environmental impact assessment' as part of the planning application process. Mandatory for schedule 1 projects. At the discretion of the planners for schedule 2 projects.
Environmental Assessment of Plans and Programmes Regulations (2004)	Requires 'strategic environmental assessment' for 'plans' and 'programmes' adopted under the national town and country planning legislation, including plans and programmes in sectors such as transport, energy, waste management, water resource management, industry, telecommunications and tourism (but only where prepared by public authorities).
The Wildlife and Countryside Act (1981) as amended by the Countryside and Rights of Way Act (2000)	Establishes powers of the countryside bodies and local authorities to establish protected species, protected areas (sites of special scientific interest, national nature reserves) and tree protection orders. All of which aim to ensure that protected trees, species and areas are only disturbed on approval by regulator.

Go to the Natural England website (www.naturalengland.org.uk) and use the map search functions to locate a Site of Special Scientific Interest near to your place of work, your home or a nature area with which you are familiar. Use the website features to find out why the site is classified and what the current conservation status is.

Using the same site find an example of protected species in each of the following categories:

A bird

A mammal

An amphibian

A plant

5.1.9 Hazardous substances

International agreements

Agreements such as the Montreal Protocol and Basel convention have been discussed in the sections on air pollution and waste. Another agreement worth of inclusion is the **Stockholm Convention on Persistent Organic Pollutants (2001)** which aims to eliminate or restrict the production and use of persistent organic pollutants (POPs). Under the convention initially twelve chemicals were either banned or highly restricted in usage including DDT and several other pesticides plus polychlorinated biphenyls (PCBs). A second group of nine chemicals were added in 2010 and there is potential for further additions as required.

European legislation

The REACH Regulation (1907/2006) imposes wide reaching requirements on the registration, testing and information provisions relating to the supply of hazardous substances within the European Union.

EC Regulation No 1272/2008 on the Classification Labelling and Packaging of Chemicals – this is the so called CLP Regulation which introduces a standard classification system for products and wastes across the EU. It is used by those selling chemicals but also in hazardous waste classification.

UK legislation

Table 5.10 Key UK hazardous substances legislation (author)

	Key requirements/summary
Control of Pollution (Oil Storage) (England) Regulations (2001)	Apply to anyone storing more than 200 litres of oil above ground at an industrial, commercial or institutional site, or more than 3500 litres at a domestic site. Require specified standards of secondary containment for storage vessels and ancillary equipment.

	Key requirements/summary
Control of Major Accident Hazards Regulations (1999)	Provide a list of tier I and tier II sites that are required to have incident control and emergency response arrangements in place. The level of control is higher for tier I sites. Jointly administered by the Health and Safety Executive and the Environment Agency to ensure that public safety and environmental protection is considered in the management of risk on high hazard sites.
REACH Enforcement Regulations (2008)	Implement the EC REACH Regulation in the UK requiring manufacturers and importers to conduct extensive hazard testing and customer information compilation. Impacts users of chemicals through phase out requirements on high hazard list and the need to be aware of inadvertent 'importer' status.

5.1.10 Civil liability and key case law

Civil actions can only be brought by the person who has been 'wronged'. Remedies that can be awarded in relation to civil claims fall under two headings:

- **Damages** – sums of money to compensate for the suffering or loss incurred
- **Injunctions** – a court order requiring those responsible for the wrongdoing to carry out remedial works or refrain from the activity causing the 'damage'

At the centre of any civil liability case is the requirement to prove that:

- a loss or harm has occurred
- a causal link exists between the loss/harm and the 'accused party'
- some kind of liability (a responsibility or in legal terms a 'tort') exists between the accused.

Proof of liability

Tort of trespass – involves direct interference and presence on property. Example potential application – fly tipping.
Tort of nuisance – disruption to reasonable expectations to use property. Example potential application – noise causing night time sleep disturbance.
Tort of negligence – no property link required. Harm or loss occurs as a result of careless act or omission which could be reasonably foreseen to cause harm. Example potential application – a fuel spill from an unbunded tank causes a road traffic accident.

Table 5.11 Nuisance considerations and defences (author)

Nuisance tests/factors which must be considered	Defences against claims of nuisance
• character of neighbourhood • standard of comfort • time and duration • motive • interference • reasonableness	• nuisance was an 'act of god' • nuisance was the 'act of a trespasser' • the act causing the nuisance took place with the consent of the plaintiff • ignorance of causing nuisance • the defence of prescription, i.e. the acts have been taking place for 20 years without complaint

Table 5.12 Key environmental civil cases (author)

Case	Key points/conclusions
Rylands v Fletcher (1867)	Established the concept of strict liability – imposing responsibility on those creating situations of danger or hazard.
Cambridge Water Company v Eastern Countries Leather (1994)	Amended strict liability to include the requirement for foreseeability but also clarified that the storage or use of polluting matter should always be considered a non-natural land use.
Hancock and Margerson v JW Roberts Ltd (1996)	Again emphasised the need for foreseeability in strict liability and negligence claims.

Case	Key points/conclusions
Hunter & Others v Canary Wharf Ltd and Hunter & Others v London Docklands Corporation (1997)	Emphasised the requirement for property rights to exist for a nuisance claim to be accepted. Also that as a general rule, a person is entitled to build on his or her own land (as long as they are in possession of statutory permissions), even if the presence of that building may interfere with his or her neighbour's enjoyment of his or her land.
Dennis v Ministry of Defence (2003)	The issue of immunity for operations that may be considered in the public interest has been overturned in nuisance claims, but the automatic right of injunction for a successful nuisance claim is removed. It also reinforced the argument that any significantly polluting activity cannot be considered a 'natural use of land'.
Empress Car Company (Abertillery) Limited v. National Rivers Authority – House of Lords ruling, 1998	Not a civil case but a House of Lords ruling on a criminal prosecution made under the Water Resources Act, 1991. The decision makes it clear that companies should address tampering and vandalism as a risk when devising environmental management systems to prevent pollution occurring from their premises. An escape, even when caused by the act of a third person, may well be a breach of duty (as vandalism should be considered a 'fact of life').

5.2 IEMA self-test questions

Question 1

You are the environmental manager for a company that produces 800kg of hazardous waste per year.

a) Identify the key legislation for the control of hazardous waste. [2]
b) Identify the regulator responsible for enforcement of this legislation. [1]
c) Describe the main actions that the company must take to comply with this legislation. [9]

Question 2

a) Identify the regulator for each of the following: [8]

Scenario	Regulator
Control of hazardous waste	
Discharge of trade effluent to sewer	
Statutory nuisance	
Registration of waste carriers	
Registration of packaging waste compliance schemes	
Discharge of effluent to controlled waters	
Permitting A2 installations	
Inspection of contaminated land	

b) Apart from the enforcement of legislation, describe the roles that regulatory bodies fulfil. [4]

Question 3

a) Outline the actions that an organisation should undertake to ensure compliance with the Producer Responsibility (Packaging Waste) Regulations 2010. [10]

b) Identify the regulator for these Regulations. [2]

5.3 NEBOSH self-test questions

Question 1

As the environmental manager of a rural food preparation and packaging business you have been made aware by staff of a waste disposal activity relating to demolition waste that was left following construction work at the site. The material has been used to infill a marshy area on adjacent agricultural land owned by a neighbouring farmer. The demolition waste includes brick rubble, soil, wood, plaster and broken asbestos cement guttering.

Outline the potential liabilities for the land owner and your own business arising from the deposit of the waste. [20]

Question 2

A paint manufacturing company wishes to discharge trade effluent to a public 'foul' sewer. In advising such a company of their obligations, describe the key points you would make under the following headings:

a) the potential advantages of pre-treating the effluent on site prior to discharge to the sewer; [7]
b) how a consent to discharge is obtained; [9]
c) the actions that may be taken, and the penalties that could be imposed, if consent limits are exceeded. [4]

Question 3

A chemical manufacturing organisation operates a Part A(1) process under the Environmental Permitting (England & Wales) Regulations under a permit issued by the Environment Agency.

a) Outline the enforcement options open to the regulator in the case of non-compliance with the conditions of the permit. (15)
b) Explain how the organisation would proceed and what they would need to demonstrate to the regulator if they wish to surrender their permit. (5)

5.4 Further information

	Further information	**Web links (if relevant)**
The Environment Agency	The Environment Agency publishes detailed guidance and technical information regarding many aspects of pollution control legislation falling under their remit. Much of this information can be found on their website.	www.environment-agency.gov.uk
GOV.UK	This website is a UK government information hub service and the business pages contain information and links relating to a variety of environmental and in particular waste related information. It also links to source legislation held on the National Archives website at www.legisation.gov.uk. In Scotland and Northern	www.gov.uk

	Further information	Web links (if relevant)
	Ireland the Netregs service (www.netregs.org.uk) provides accessible summaries of legal requirements organised by issue and industrial sector – an excellent resource.	
WRAP (Waste and Resources Action Programme)	Government funded, best practice programme providing extensive best practice and case study information	www.wrap.org.uk
European Environmental Agency	European Directives and policy	www.eea.eu.int
The Department of Environment, Food and Rural Affairs (DEFRA)	Maintain an excellent website covering legislation, policy and consultation documents. DEFRA also plans to launch a new environmental legislation hub which will be known as DEFRA-LEX but at the time of writing no definite timescales could be confirmed.	www.gov.uk/defra
Institute of Environmental Management and Assessment	In particular the legal brief section of the Environmentalist magazine	www.iema.net
The Joint Nature Conservation Committee	Information on UK Special Protection Areas (SPAs) and Special Areas of Conservation	jnccdefra.gov.uk
Magic website	This is an interactive website maintained by Natural England that provides information on all sites of conservation interest including SSSI's across the UK.	www.natureonthemap.naturalengland.org.uk
Commercial Organisations	There are a number of commercial organisations which will provide information about current environmental issues for a subscription fee. Examples of the more popular bulletins include: the Ends report; McCormick information bulletins; Technical Indices; Barbour Index; Croner's Environmental Management and Case Law; GEE's Environmental Compliance Manual.	

CHAPTER 6

Understanding environmental management and sustainable development in a business context

This chapter is organised into the following sections:

6.1 Revision notes

6.1.1 How do organisations impact the environment

Direct impacts

These are environmental issues that occur as a result of activities performed by the organisation itself. The group includes issues such as pollution impacts and impacts related to the consumption of renewable and non-renewable resources.

Supply chain impacts

These are environmental issues that are caused by third parties who are carrying out activities linked to or supporting the organisation's activities (i.e. suppliers and contractors) and might include transport impacts associated with the supply of products and services.

Product/service impacts

These are impacts that are caused by customers or others using the product produced or as a result of the service supplied by the organisation. An example could be packaging waste disposal by customers.

6.1.2 Pressures driving corporate environmental performance

- Legal – compliance with ever increasing legislation
- Financial – resource availability, plus pressures from insurers and financial institutions

- Market – customer demands, reputation and profitability associated with resource efficiency
- Social – local community and general public perception as well as lobby groups and the press

Such pressures create potential for business opportunities as well as threats.

6.1.3 Managing sustainability threats and opportunities

Sustainability threats

- Raw material availability and increasing cost driven by non-renewable resources becoming scarcer, e.g. fossil fuel based materials, or through regulation of supplies to ensure the most socially effective use of non-renewables. In some instances increased standards of management of renewable resources may also lead to an end of 'cheap goods' where prices do not currently reflect the environmental cost of the materials, e.g. palm oil, low grade plastics etc.
- Escalating energy and transport costs and availability – as with raw materials, energy cost is likely to escalate in the short term, in particular through a combination of decreasing availability of fossil fuels and government imposed financial incentives to increase efficiency and drive the shift to more renewable sources of energy.
- Increased legal constraints are likely to be part of the education – taxation – regulation package of measures used by government to drive the shift to sustainability. Whether related to eco-efficiency, product design, supply chain management, material resourcing or consumer information we can expect to see a continued flow of legal pressures pushing all organisations and especially stragglers in the move towards sustainability to keep pace with the best in class innovators.
- Market share threats is another looming issue with consumer awareness showing a steady (if somewhat patchy) increase

over the past twenty years. Organisations that fail to identify and respond to shifting priorities or sustainability concerns amongst their customers may end up losing ground to competitors.

Sustainability opportunities

- As many of the pressures around sustainability relate to inefficiencies in material and energy usage, organisations that respond well may not simply avoid the disadvantages of escalating costs but make savings through becoming increasingly efficient and streamlined throughout their direct and indirect operations.
- Organisations that position themselves and their products and service well in terms of key reputation measures are likely to be able to turn such trends to their advantage in terms of market share.
- The long term perspective demanded by sustainability planning may provide organisations with the vision and wherewithal to adapt and secure their long term viability in a way that more short sighted organisations may struggle to achieve.

Task

Prepare a short presentation to a senior management team on what you see as the key sustainability threats and opportunities facing your organisation over the next ten years.

A value chain perspective

This really just means seeing the wider context in which the organisation operates when considering sustainability threats and opportunities (see Figure 6.1).

A risk management approach

Hazard – a condition or situation with the potential for an undesirable consequence (akin to an environmental aspect in ISO14001 terminology).

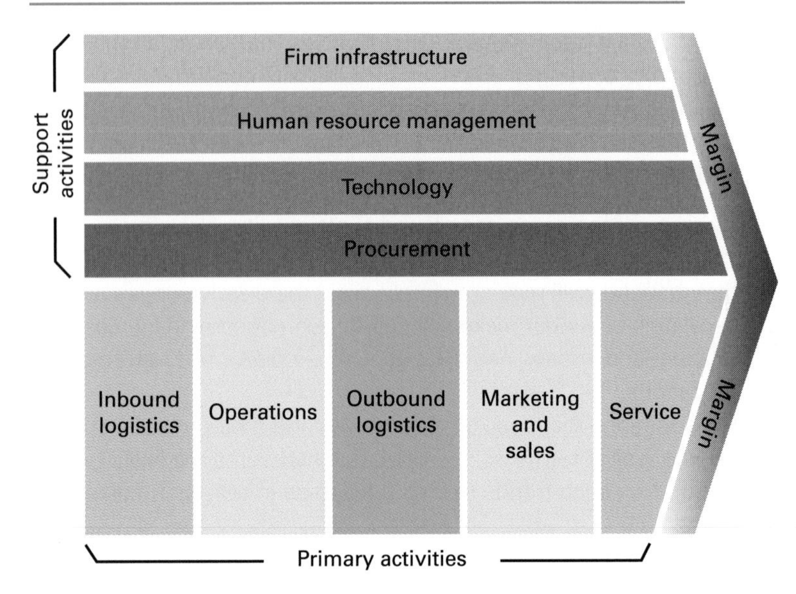

Figure 6.1 Porter's value chain

Source: Image by permission of D.P. Singh via Wikimedia Commons

Harm – the adverse consequences resulting from the realisation of a hazard (akin to the environmental impact in ISO14001 terminology).

Risk – the potential for the realisation of undesirable consequences, i.e. a combination of the likelihood and consequences of a specific outcome.

The basic risk management model:

1. Identify and prioritise risks
2. Minimise and mitigate risks
3. Monitor and provide assurance that hazards are adequately controlled

It should be noted that, while the risk management approach tends to be familiar and therefore acceptable to many people it is in essence a 'defensive' strategy in relation to organisation change. In the mid to long term transitioning to a more sustainable organisation will demand a more proactive and 'opportunity seeking' approach.

6.1.4 Becoming a more sustainable organisation

It may be useful to consider four evolving stages of involvement in environmental management and sustainable development:

Stage 1 – being reactive to issues

Stage 2 – developing management systems

Stage 3 – adopting an integrated strategic approach

Stage 4 – organisational transformation leading to the sustainable enterprise

Driven by the threats and opportunities discussed above organisations are responding in three key areas as they move through stages 3 and 4:

- **Increasing eco-efficiency** – reducing the resource intensity of their products/services
- **Managing their supply chain issues** – through specification, partnerships or selection strategies
- **Improving stakeholder communications** – product/service information, company performance information, consultations processes

Table 6.1 Stages of sustainability in organisations (author)

	Characteristics	**Drivers**
Stage 1 – reactive	A fire fighting approach – dealing with issues as they are brought to our attention by 3rd parties, e.g. complaints, legal breaches or demands	External stakeholder's interests or the threat of prosecution
Stage 2 – developing environmental management systems (EMS)	A more coordinated and voluntary approach to control and improvement of environmental issues. Often involving ISO 14001 certification etc.	Reactive – customer specification, regulator requirement Proactive – reputation enhancement, resource efficiency objectives

	Characteristics	Drivers
Stage 3 – adopting a strategic approach	Established EMS in place and focusing on impact reduction not simply control measures. Some redesign of process and product/service to reduce impacts. Developing a longer term view of improvements that cover infrastructure, product and supply chain change rather than simply 'tweaking' what is already present	Reactive – customer specification, regulator requirement Proactive – reputation enhancement, resource efficiency objectives
Stage 4 – sustainable enterprise	A major shift in perspective with full life cycle thinking built into corporate decision making at all levels – short, medium and long term. Organisations won't be wholly sustainable today but they will have a clear view of how they intend to move towards this goal and will be communicating with all key stakeholders to develop understanding and cooperation. In some instances companies will be considering fundamental changes to the nature of their products, services, business structure in order to facilitate such change	Proactive – a high level recognition that sustainability means short term efficiency, mid-term market gain and long term business survival

6.2 IEMA self-test questions

Question 1

a) Identify the range of risks that an organisation may face in relation to environmental issues. [6]
b) Outline the measures that could be used to manage these risks. [6]

Question 2

Outline the business benefits of taking a proactive approach to environmental management. [12]

Question 3

a) Explain what is meant by sustainable development. [4]
b) Identify 8 factors that a supermarket may consider in its sustainable development strategy. [8]

6.3 NEBOSH self-test questions

Question 1

Describe the business reasons for effectively managing environmental risk. [20]

6.4 Further information

Topic area	Further information sources	Web links (if relevant)
Risk management	IEMA practitioner guide, 2006. Risk management for the environmental practitioner. IEMA	www.iema.net
Eco-efficiency	World Business Council for Sustainable Development	www.wbcsd.org
Sustainability strategy	Information on a company transformed by commitments to sustainability	www.interfaceglobal.com/sustainability

CHAPTER 7

Collecting, analysing and reporting on environmental information and data

This chapter is organised into the following sections:

7.1 Revision notes

7.1.1 Types and sources of environmental data/information

A definition of data – 'facts and statistics collected together for analysis'.

A definition of information – 'that which is conveyed or represented by a particular arrangement or sequence of things'.

Measurements made at an effluent discharge point provide **data** in the form of readings in mg/l of specified pollutants. Collation of the data into peak readings, daily averages and calculated monthly emission totals provides us with ***information*** that we can compare with limits set by the regulator to determine our compliance status with permit conditions.

Reasons for sourcing and monitoring a wide variety of data and information:

- To inform decisions about environmental priorities, i.e. to tell us what our issues are.
- To track performance over a period of time, i.e. to tell us whether our efforts are getting the results we hope for.

Performance indicators (ISO14031)

Environmental performance indicators (EPIs):

- Management performance indicators – measure organisational effort, e.g. number of audits conducted per month.
- Operational performance indicators – measure results, e.g. tonnes of wastes generated per year.

Environmental condition indicators (ECIs) – measure the state of the receiving environment, e.g. ambient air quality.

Performance indicators selected to track progress should reflect key aspects and stakeholder concerns.

7.1.2 Data selection principles

Choosing what to measure and how to present the data we collect may be an important step in ensuring we generate information that will be useful in guiding decision making and/or tracking progress.

Qualitative vs quantitative data

Qualitative data may be defined as information based on opinion, judgment or interpretation, e.g. perceived nuisance potential based on detectable levels of noise or odour, for example at a site boundary.

Quantitative data may be defined as numerically measured criteria that can be subject to statistical analysis and/or benchmarking, e.g. air emissions measured at a discharge point in milligrams per cubic metre (mg/m3).

Quantitative data is generally preferred but where impacts are qualitative, e.g. nuisance, visual impacts etc. it is important to consider qualitative assessment too.

Pollutant vs resource usage data

At its simplest this is measuring inputs (resources) and outputs (pollutants). In practice, however, pollutant monitoring often involves some calculations based on inputs as well as monitoring of emissions, discharge or wastes. Similarly while resource usage is often focused on inventories and purchasing data, often some kind of break down monitoring is required to understand patterns of usage and identify opportunities for savings.

Normalised vs absolute data

Absolute data is the actual amount of something measured, e.g. total tonnes of waste generated by an organisation in a day/month/year. Absolute data gives the truest picture of the scale of impact created by the organisation, however, it can be misleading in judging levels of control or improvement over time if there are multiple variables involved.

Normalised data is based on absolute data but is subject to some kind of referencing process to enable comparison between different locations or over time. An example might be tonnes of waste generated per unit of product produced.

AA1000AS (2008)

Sets out three basic principles that should be applied when selecting performance monitoring data within an environmental or sustainability programme:

- **Inclusivity** – performance data and reporting should relate to the interests of key stakeholders.
- **Materiality** – data should be relevant and presented in such a way as to be a true representation of impact, avoiding misleading or 'glossy interpretations' that present information in a way biased towards the organisation. Associated with the principle of materiality is the concept of 'completeness', i.e. that all material issues should be covered and reported.
- **Responsiveness** – demonstrating reactivity to stakeholder concerns and meeting their information needs.

So in essence AA1000 is saying any organisation when using data and information as part of its environmental programme should:

- Consider who and what will be impacted by its activities
- Monitor those things relevant to such impacts and process information in a complete and unbiased way
- Report such information in an appropriate way to all key stakeholders.

Identify some appropriate key performance indicators for your own organisation that will be useful to track performance in the medium to long term (say 10–15 years). To do this you will have to consider the potential changes in scale or nature of company operations, as well as the potential improvements that are likely to be made in dealing with key environmental impacts. Your indicators should still be valid even with significant changes in both areas.

7.1.3 Verification and assurance

Depending on the degree of reliability and credibility required one or both of the following may be relevant:

Calibration and corroboration of data sources

Ensuring equipment remains within required standards of accuracy. Also using regular cross checks to corroborate findings and ensure that main data source is continuing to be reliable.

Internal and independent auditing/validation of data

Internal audit programmes, independent data verification as part of EMAS or AA1000AS accreditation or as part of the Environment Agency's 'Operator Monitoring Assessment' process – all are checking the accuracy and reliability of data collection, compilation and reporting processes. The degree of validation required varies but at its most rigorous, assurance auditors will evaluate everything from calibration of equipment through competency of personnel, sampling of data sets produced at ground level and management and calculation processes used to generate corporate performance figures.

7.1.4 Third party benchmarking

Why – to enable organisations to see how they are doing, and to enable stakeholders to compare the performance of competitors/ peers.

Three examples appear to be leading the pack at present – the Dow Jones Sustainability Index (DJSI), the Business in the Community (BITC) Corporate Responsibility Index and the FTSE4Good index. All are based on independent reviews of questionnaires completed and data provided by participating companies.

7.1.5 Pollution and nuisance monitoring

Measurement – the quantification of pollutants achieved by some kind of gauging.
Monitoring – the collection and interpretation of a number of measurements or estimates over a period of time. Often monitoring involves some kind of comparison with a reference standard.
Key guidance on appropriate standards for equipment, personnel competency and sampling and analysis protocols are provided by the Environment Agency under the Environmental Permitting horizontal and technical guidance notes, the Monitoring Certification Scheme (MCERTS) and the Operator Monitoring Assessment (OMA) guidance note.

Analytical techniques

Table 7.1 An overview of analytical techniques used in pollution monitoring

Type of analysis	Basic principle	Examples of analytical techniques/ pollutants
Chromatography	Separates pollutants in a mixture allowing them to be quantified individually by another technique	Ion chromatography, gas chromatography, liquid chromatography. Used for the separation of a wide variety of pollutants including pesticides, chlorinated solvents, polychlorinated biphenyls, dioxins and endocrine disruptors, VOCs etc.
Electrochemical	Measures electrical properties related to chemical composition	Ion selective electrodes (including pH meters) and conductivity meters are used for individual pollutants, e.g. metals and total salts (i.e. combined pollutant load)
Gravimetric analysis	Measures the mass of pollutant present	Filtration techniques, e.g. total solids and suspended solids in water; PM_{10} and total suspended particles in air
Optical	Assess optical properties of a sample either qualitatively	Colour, turbidity (of water samples), obscuration, opacity (for particulates samples)

Type of analysis	Basic principle	Examples of analytical techniques/ pollutants
	(e.g. colour) or quantitatively (e.g. via light transfer)	Visible oil and grease
Spectrometry	Techniques involve energy from different parts of the spectrum. In all cases, from radiation absorbed or emitted, information is obtained on the composition of the sample and the amounts of constituent pollutants	Atomic absorption spectrometry (used for heavy metals) Chemiluminescence analysis (used for oxides of nitrogen in air) Infrared spectrometry (used for SO_2, total hydrocarbons) Mass spectrometry (often coupled with gas chromatography and used for many organic pollutants, e.g. PCBs, solvents etc.)
Volumetric	Measures the volume of one (known) substance reacting at a fixed ratio enabling the amount of the other chemical (unknown) to be inferred	Titrations for pH – generic chemical properties – exact pollutant unspecified Sorbent tubes and diffusion tubes – e.g. Draeger tubes (a particular brand name). These glass tubes contain a pollutant specific reagent adsorbed onto an inert solid. A fixed volume of gas is drawn through the tube using a hand pump. The sample time is a few seconds, during which (if the pollutant is present) a colour develops from the sampling end of the tube. At the end of the sampling period, the colour should extend along a fraction of the length of the tube. The tubes are pre-calibrated with a concentration scale, so that the distance the colour has travelled can be directly related to the gas concentration. This technique can be applied to a wide variety of pollutants by varying the reagents in the tubes.

Source: Amended from Brady et al., 2011. *Environmental Management in Organisations: the IEMA Handbook*, 2nd ed. Earthscan.

Monitoring strategies

Source monitoring vs ambient monitoring – depending on goal of programme, and range of influences in the receiving environment.
Typically a monitoring programme will involve one or more of the following:

Continuous monitoring – measurements made continuously with few or no gaps in data collection.
Periodic monitoring – measurements made at defined intervals and/or under defined operating conditions.
Surrogate monitoring – the pollutant itself is not measured but is estimated from another parameter, e.g. fugitive emissions of VOCs estimated from consumption totals of cleaning solvents.
In all cases sampling should be carefully controlled to ensure that samples are representative of temporal and spatial variation. A summary of monitoring considerations is provided in Table 7.2.

Table 7.2 Summary of monitoring considerations (author)

Monitoring air pollution	Source based or ambient air monitoring Continuous vs periodic sampling Manual vs automated monitoring Active vs passive sampling
Monitoring water pollution	Considerations as per air pollution monitoring. Classic suite of parameters: Dissolved oxygen, biochemical oxygen demand (BOD), chemical oxygen demand (COD), total solids, suspended solids, pH, ammoniacal nitrogen, electrical conductivity (measures inorganic salt content) Ecological classification of inland waters now follows a European standard introduced under the Water Framework Directive (2000).
Monitoring/ assessing land pollution	Stage 1 assessments – risk assessment audit looking at likelihood of contamination and trying to identify type and location of likely contaminants. Stage 2 assessments – intrusive sampling and analysis to determine:

	(a) whether contamination exists; (b) whether a risk is posed to human health and/or the environment; (c) whether there is a need for clean-up to mitigate such impacts; (d) the nature of on-going monitoring programmes, if required. Sampling via – surface samples, trenches, augers and boreholes. Analytical techniques depend on pollutant phase (gaseous, liquid or solid) and type.
Monitoring noise nuisance	Measurement in dB(A), i.e. linked to range of human hearing. Must consider noise properties such as reflection, propagation and annoyance factors. Nuisance tends to be highly context based, i.e. the louder a noise is above background levels the more likely it is to be a nuisance. BS7445 provides guidance on monitoring environmental noise and BS4142 is a nuisance prediction methodology.
Monitoring odour nuisance	Source vs receptor based monitoring Chemical analysis can be used for simple odours, e.g. VOC's. But sniff tests/subjective evaluation is the only practical option for complex odours such as those generated by landfill sites, sewerage treatment works etc.
Monitoring dust nuisance	Passive receptor based sampling gives the best indication of nuisance potential. Techniques include – gauges, sticky pads and slides.

7.1.6 Methods for interpreting data and information

See Chapter 9 of this study guide (Chapter 7 of the *Manual*) for details but methods include:

- Aspects evaluation and prioritisation methodologies
- Environmental impact assessment
- Comparison with legal or other standards
- Environmental modelling
- Statutory contaminated land assessment risk assessment
- Noise nuisance prediction using BS4142
- Planning guidance relating to noise nuisance

- Dust and odour nuisance prediction models
- Ecotoxicity assessments and the use of indicator species.

7.1.7 Methods for communicating data and information

There are many ways to communicate data relating to environmental performance ranging from corporate reports through to colour coding of bins or labelling of products. Keys to effective communication are:

- clear identification of target audience
- clear identification of communication goals
- tailoring of communication content and method to suit the above.

7.2 IEMA self-test questions

Question 1

You have been asked to establish a system for monitoring your organisation's waste production and disposal. Describe the data you would collect and explain how you would present this data in order to monitor performance against targets over time. [12]

Question 2

a) Explain what is meant by 'Absolute Data' and 'Normalised Data' illustrating your answer with an example of each. [6]
b) Outline the principles you would apply when disseminating data about performance. [6]

Question 3

a) Describe the role of assurance and verification in relation to environmental data. [6]
b) Explain what is meant by the principles of 'materiality', 'completeness' and 'responsiveness' in data assurance processes. [6]

7.3 NEBOSH self-test questions

Question 1

You are the environmental manager of a medium sized construction company. **Describe** the indicators that you might propose using to measure the organisation's environmental management performance. [20]

Question 2

a) Oxides of nitrogen (NOx) are an important group of air pollutants. **Describe** the main environmental impacts arising from their emission into the atmosphere. [14]
b) **Outline** two methods that could be used to quantify NOx emissions generated by an organisation. [6]

7.4 Further information

Topic area	Further information sources	Web links (if relevant)
Choosing and presenting data	ISO14031:2013 Environmental management – Environmental performance evaluation – guidelines AA1000AS (2008)	www.iso.org www.accountability.org
Assurance and verification of data	AA1000 Assurance Standard (2008) The Environment Agency Operator Monitoring Assessment scheme DEFRA, 2013 Environmental reporting guidelines – including mandatory greenhouse gas reporting guidance. DEFRA	www.accountability.org www.environment-agency.gov.uk www.gov.uk/defra
Data collection and analysis	Brady et al., 2011. *Environmental Management in Organisations – the IEMA Handbook*, 2nd ed. Earthscan Reeve, R., 2002. *Introduction to Environmental Analysis.* John Wiley & Sons Ltd BS7445 – Description and measurement of environmental noise	www.iema.net

CHAPTER 8

Environmental management and assessment tools

This chapter is organised into the following sections:

8.1 Revision notes

8.1.1 Aspects identification and environmental risk assessment

Useful to distinguish between two groups of environmental assessment methodologies:

- 'Environmental impact assessment' relates to planned activities.
- 'Environmental audit' relates to existing activities.

However, they share some characteristics. The following can be considered a selection of tools all used to do the same job – the identification and prioritisation of environmental issues:

- Aspects checklist
- Input–output analysis (an activity based approach)
- Consideration of normal, abnormal and emergency operating conditions
- Consideration of source – pathway – receptor
- Significance assessment methodologies (these are typically different between planned and existing activities).

Aspects checklist:

- Hazardous material usage
- Air emissions
- Effluent discharges
- Solid waste
- Raw material usage
- Energy usage
- Land usage
- Nuisance

Input-output analysis

Identify all functions in an operational area and consider inputs and outputs to those functional units (these correlate to the aspects checklist categories). See Table 8.1.

Table 8.1 Input–output analysis (author)

INPUTS	ACTIVITIES	OUTPUTS
Typically: • Raw materials • Components • Water • Energy • Packaging • Hazardous materials	For example: • Production processes • Materials storage and handling • Product design • Transport and distribution • Office/administration functions	Typically: • Products • Packaging • Wastages • Effluent • Air emissions • Heat • Noise • Odour

Different operating conditions

Consider normal, abnormal and emergency operating conditions – each may have very different environmental aspects. Need to clearly define abnormal and emergency conditions to enable frequency and likelihood to be considered when assessing significance.

Source-pathway-receptor linkages (see Figure 8.1)

EIA often starts from receptor and works back to aspect. Audits normally work the other way round but it can be useful to consider what key receptors exist locally to highlight potential or actual issues that may affect them.

EIA risk assessment

Always done for proposed developments and always conducted by 'experts' in the relevant impact/receptor fields (archaeologists, ecologists, etc.).

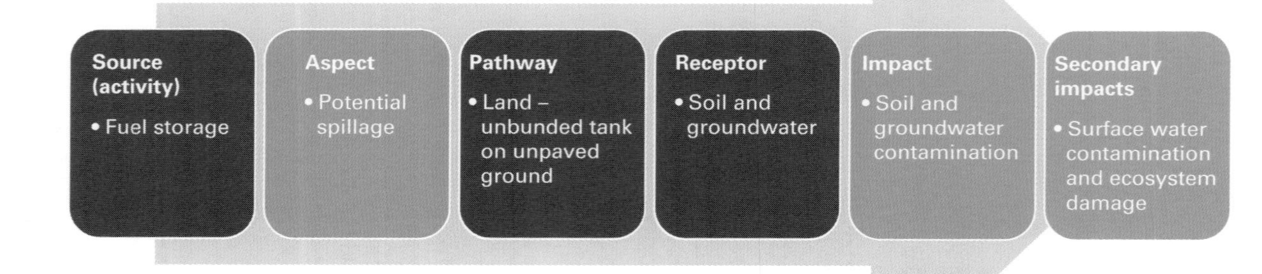

Figure 8.1 Source – pathway – receptor linkages

It is based on consideration of receptor survey and modelling data. Findings are peer reviewed by regulators and statutory consultees.

EMS/corporate environmental risk assessment

Definitions:

Hazard – a condition or situation with the potential for an undesirable consequence (akin to the environmental ***aspects*** in ISO14001 terminology).

Risk – the potential for the realisation of undesirable consequences, i.e. a combination of the likelihood and consequences of a specific outcome.

Harm – the adverse consequences resulting from the realisation of a hazard (akin to the environmental **impact** in ISO14001 terminology).

Caution – the terminology can tempt you to only consider 'emergency scenarios' and 'pollution impacts'. We know by now that these are only part of the issue so it is important to remember that 'environmental risk assessment' must adopt a wider view than traditional health and safety focused risk assessment.

Assessing significance/prioritising risks

Purpose:

- To identify those aspects (hazards) that should be the focus of our attention in the EMS both in terms of controls and improvement initiatives.
- To enable a consistent approach to prioritisation over time and in different parts of the organisation.
- To make transparent our approach to prioritisation so that stakeholders (including the certification bodies) can see the logic and criteria we have used.

Process – significance assessment in organisations (i.e. the identification and prioritisation of risks/aspects) is normally a non-scientific evaluation based on business and stakeholder priorities.

Clarity and consistency of decision-making is critical. Approaches vary widely between organisations but all methodologies should achieve the characteristics shown in Figure 8.2.

Formats of methodologies – classic formats for risk assessment/significance evaluation methodologies include matrices, flow charts and questionnaires. Results are typically presented in the form of a table or register.

8.1.2 Environmental auditing

Audit scope and purpose varies widely in terms of technical and organisational detail. Common audit goals, which may be undertaken in isolation or combination, include:

- Identification of risk and/or opportunity and the development of associated action plans
- Provision of documented assurance of compliance with own or external standards
- Provision of best practice advice by the auditor to the auditees.

Standard audit steps (applicable to any audit type):

1. Set scope – identification of objectives, time/geographical/organisational limits
2. Data review – documentary information of some sort
3. Site investigation – visual inspection plus interviews and further data review
4. Assessment of findings – against some standard or reference (variable depending on scope)
5. Report compilation – verbal and/or written, style, length and content dependent on audience and scope of audit.

Examples of different audit types

- EMS internal audits
- EMS certification audits
- Duty of care audits
- Supply chain audits
- Preliminary environmental review
- Waste reviews

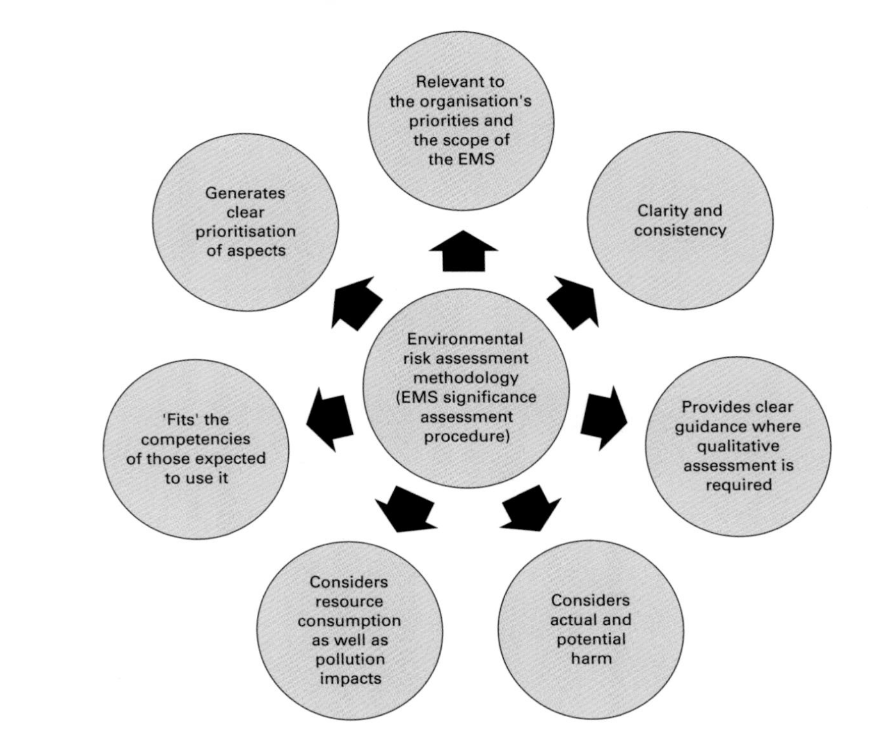

Figure 8.2 Risk assessment methodology criteria

- Energy audits
- Contaminated land assessments
- Due diligence/pre-acquisition audits.

8.1.3 Environmental management systems

Potential benefits of a 'systems approach' to environmental management

Licence to operate

- Assist in obtaining authorisations, consents and permits.
- Avoid enforcement action.
- Improve business and the regulatory relationship.
- Maintain and enhance community relations.

Market size and share

- Assure customers of commitment to responsible environmental practices.
- Identify customer concerns and opportunities for developing and sharing environmental solutions.
- Meet or exceed vendor certification criteria.
- Develop products and services which will improve the environmental performance.
- Retain and/or improve market share.
- Create new products and new market opportunities.

Cost control

- Improve process and operational efficiency.
- On-going savings in materials, energy and waste costs.
- Avoid the liability costs through criminal proceedings and/or civil actions.

Access to financial services

- Satisfy investor criteria.
- Obtain insurance at reasonable costs.

Public image

- Enhance public image.
- Demonstrate responsible care.

Potential disadvantages of a 'systems approach'

- Cost, bureaucracy, and ongoing commitment.

Two principal models of EMS

Now closely aligned – ISO 14001 and the EC Eco Management and Audit Scheme (EMAS).

ISO 14001 is one of a series of standards developed by International Standards Organisation. EMAS is an internationally used scheme now aligned to ISO14001.

Key difference is EMAS requirement for a verified public statement that includes specific performance data, e.g. tonnes of waste produced, whose accuracy is audited by the verification body.

EMS implementation

EMS implementation does not normally follow ISO 14001 clause by clause. More typically an 'assess, plan, implement and operate' staged approach is used.

It should be noted that there is a high degree in flexibility in the clauses of the standard and this allows significant variation in the exact details of the management system between different organisations. This is considered a strength of the approach as each organisation can develop a system that 'fits' its culture, existing business systems and level of environmental risk.

The certification process

ISO 14001 certification and EMAS verification audits are carried out by registered organisations, following standard guidance.

Typically a 2 stage process – initial assessment and main audit.

Management system integration is generally seen as beneficial and desirable. Although no single certification is available for integrated management systems many certification bodies have procedures to combine/streamline certification processes.

Staged certification schemes have been developed that allow recognition at different stages of implementation. Schemes include BS 8555, ISO 14005, IEMA's Project Acorn and Groundwork Wales' Green Dragon programme.

8.1.4 Resource efficiency

The first step in any resource efficiency programme is some kind of survey or investigation to build understanding of resource usage, waste generation points and hence identify opportunities for improvement.

Materials management to reduce impact

Essentially there are 3 core strategies:

- Use less materials – e.g. avoid overordering.
- Use materials from lower impact suppliers – e.g. buy locally or from ISO14001 certified companies.
- Replace materials with high pollution potential with lower pollution potential alternatives, e.g. solvent free inks, paints etc.

Managing waste to reduce impact

Based on the principles of the waste hierarchy. Elimination/reduction is covered in materials management so here the focus is on:

- reuse, e.g. returnable packaging
- recovery, e.g. recycling and composting initiatives.

Energy (carbon) management

Based on improvements in:

- consumption efficiency – switch off campaigns
- transport efficiency – fuel type/vehicle selection
- supply efficiency – CHP usage
- switch to renewables/lower impact energy sources – green tariff or self generation.

8.1.5 Environmental footprinting

A technique used to communicate the overall impact of an organisation, country or indeed the whole of humanity in an easily understood and graphic way. Essentially an environmental footprint is the land area calculated as necessary to supply the resources consumed and absorb the wastes generated by an individual, organisation or population.

Approach promoted by WWF and linked to an annual global report and a set of guidelines and commitments entitled 'one planet living'.

Task

Go to the 'www.myfootprint.org' website. Complete the online calculator to see what your ecological footprint is and how it compares with the UK and other international averages. As you answer the questions think about how the programme might work and what assumptions might be included to calculate the footprint area.

8.1.6 Environmental impact assessment

EIA is a planning tool that examines the environmental consequences of development actions in order to facilitate decision making and the identification of avoidance and mitigation measures. The process is conducted by a group of specialists and comprises the following steps:

1. project description and definition of assessment area
2. impact scoping (often including consultation with key stakeholders)
3. baseline environmental descriptions/surveys
4. identification of key aspects
5. impact assessment
6. mitigation proposals (iterative loop back to step 4)
7. residual impacts
8. consideration of cumulative impacts (from other projects)
9. preparation of environmental statement (and submission as part of planning application).

Good practice ensures clarity of links between EIA process and the life cycle of the development including transfer of information and commitments into construction, operation and decommissioning phases as appropriate.

Task

Your organisation is planning to build a major new facility, comprising a large throughput, solvents based printing process. As part of the development, an effluent treatment plant and an on-site power generation facility will need to be built. The selected site is adjacent to an urban area on the bank of a river that is noted for its fish stocks and conservation interest.

Consider the project in two stages – construction and operation – and then list what you would consider to be the key aspects/ potential impacts that would need to be considered in an EIA conducted as part of the planning process.

Construction stage issues	Operation stage issues

8.1.7 Strategic environmental assessment

Applies the EIA preventative approach to the development of (public) plans and programmes. Like EIA this is a statutory requirement arising from EU legislation. Although assessments obviously have to be done at a 'higher level' essentially the process is the same as for EIA with public consultation and incorporation of feedback as the final stages. SEA is required in relation to the following sectors:

- Agriculture
- Forestry
- Fisheries
- Energy
- Industry
- Transport
- Waste
- Telecommunications
- Tourism
- Planning
- Land use

8.1.8 Life cycle assessment and eco-design

A common code of practice exists for conducting a life cycle assessment (ISO14040 series) and incorporating eco-design into environmental management systems (ISO14006).

Life cycle – The consecutive and interlinked stages of a product system, from raw material acquisition or generation of natural resources to final product disposal.

Life cycle assessment (LCA) – Compilation and evaluation of the inputs, outputs and the potential environmental impacts of a product system throughout its life cycle.

Eco-design – a method of systematically integrating environmental considerations into the design of products, processes and services throughout their life.

LCA assesses the different environmental issues that occur over time in different geographical sites and at the different stages a product will pass through in its lifetime. By taking such a wide view,

LCA aims to avoid the problem often encountered in environmental improvement programmes and eco-design initiatives of shifting impacts from one part of the life cycle to another.

ISO 14040 identifies 4 stages within the LCA process:

1. definition of goal and scope
2. inventory analysis
3. impact assessment
4. interpretation of results.

Streamlined LCA concentrates solely on the stage(s) or burdens identified as important for the product by the earlier LCA(s).

Table 8.2 LCA applications and benefits (author)

Application	Benefits
Improved efficiency and cost savings	Tracking energy and material inputs as well as waste outputs through production can identify opportunities for efficiency improvements and hence, cost savings.
Product design	As an aid to product design decisions, LCA can help identify the environmental pros and cons of different options. Such information may avoid decisions that could have long term implications in terms of producer responsibility or simply guide design to maximise benefits of 'green marketing'.
Product marketing	In order to promote the environmental performance of a product with confidence and credibility, claims need to be based on concrete evidence. LCA, particularly where product comparison has been undertaken, is well suited to this function.
Supply chain pressure	For businesses such as retailers, the primary environmental impacts are upstream (with suppliers) or downstream (with consumers). Such organisations are becoming more interested in reducing their environmental impacts, particularly through influencing their supply chain. Suppliers able to demonstrate the environmental performance of their products through LCA may be well placed to win increasing market share with such organisations.
Informs procurement decisions	This is essentially the flip side of the previous item – LCA reports allow interested parties to make informed decisions in order to minimise the impact of their procurement choice.

LCA applications and benefits

Eco-design does not require a detailed LCA to be completed as a pre-requisite. A broad understanding of the key product design issues through the life cycle is adequate.

As a design philosophy, some commentators advocate consideration of a product as **'a vehicle for the provision of one or more services'**. For example – a car should not be considered as a metal box on 4 wheels powered by a combustion engine but as a unit that delivers customer service in relation to personal transport, status symbol, entertainment, extension of personal space outside the home, comfort zone etc. By considering the existing design of any product in this way we are more likely to identify new alternatives rather than become stuck in refinement of existing formats.

The eco design approach is becoming a legal requirement in some instances via legislation such as the EU Eco-design Directive (2009) which sets performance standards primarily in relation to energy using products.

8.1.9 Sustainable procurement

Sources of guidance on product and supplier selection include:

- the Mayor of London's green procurement code
- the US environmental protection agency's green procurement programme
- B&Q Quest programme
- Worldwatch institute guide
- BS8903 Principles and framework for procuring sustainably (2010).

DEFRA has made available a prioritisation methodology that is aimed at assisting organisations to decide where they should focus their efforts in order to reduce the environmental and social impacts caused by its supply chains. The methodology is presented in the form of an active spreadsheet tool with accompanying guidance. This and similar methodologies aim to select the most significant suppliers and/or products used by the organisation. These can then be the focus of such initiatives as:

- Pre-qualification questionnaires
- Selection standards using defined company performance criteria, e.g. ISO14001
- Selection standards using defined product performance criteria, e.g. timber and timber products carrying the Forestry Stewardship Council approved logo.

8.1.10 Carbon management

Carbon management programmes with or without carbon neutral goals are becoming increasingly common, driven by growing corporate awareness, fiscal measures to drive efficiency (e.g. climate change levy, EU emissions trading scheme etc), and statutory reporting requirements.

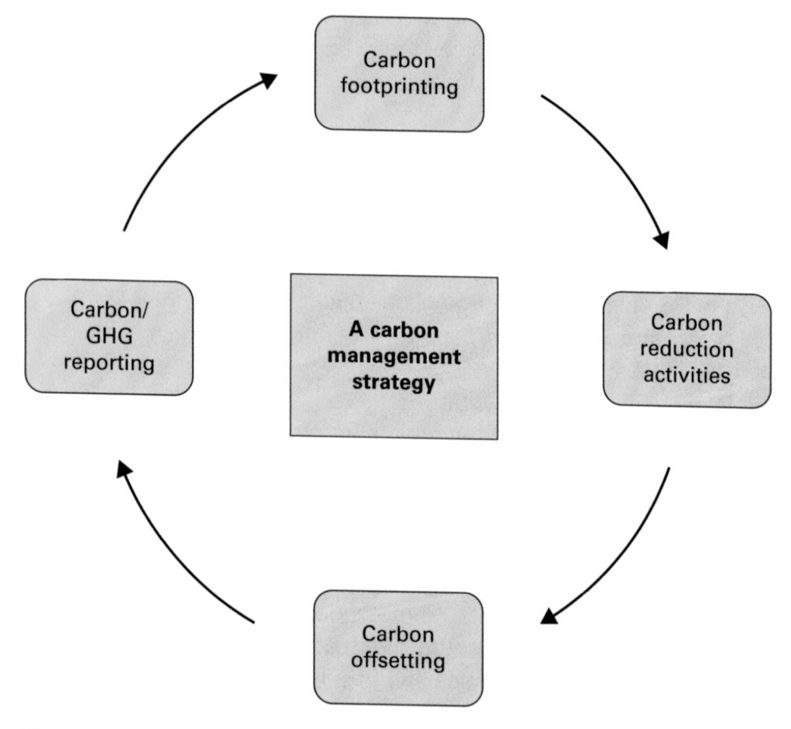

Figure 8.3 A carbon management strategy

The **carbon footprint** of any organisation comprises direct and indirect carbon dioxide emissions or, more strictly, greenhouse gas emissions. DEFRA guidance provides a standardised methodology for calculating and transparently reporting greenhouse gas emissions in a carbon dioxide equivalent format for comparative purposes.

Carbon reduction activities may encompass a broad range of initiatives ranging from efficiency improvements to transition to low carbon alternatives in relation to power, heat, light and transport.

Carbon offsetting is often about the protection, enhancement or creation of carbon sinks, e.g. forests, but may also involve investment in renewable energy in areas with expanding energy demand (thereby avoiding the creation of additional carbon intensive power generation).

Carbon management reporting is mandatory in the UK for all publically listed companies and may be rolled out to smaller organisations in coming years. DEFRA guidance ensures consistency of reporting standards and affords regulators and other stakeholders the ability to compare the carbon management efforts of an organisation over time or to make comparisons within business sectors.

The carbon disclosure project is a very successful initiative 'encouraging' the biggest companies globally to report on their greenhouse gas emissions. The project aims to generate information for investors and other stakeholders and provides further incentive for organisations to implement a carbon management programme.

Task

List the direct and indirect sources of greenhouse gases that should be considered were you to complete a carbon footprint calculation for your organisation.

8.1.11 Corporate ecosystems review

Produced by the World Resources Institute, this methodology assesses an organisation's dependence on, and impact on, a whole range of ecosystem services under 4 main headings:

- Provisioning services, e.g. crops, livestock and capture fisheries
- Regulating services, e.g. regulation of water timings and flows
- Cultural services, e.g. recreation and tourism
- Supporting services, e.g. primary production by organisms at the base of the food chain.

In 2011 the World Business Council for Sustainable Development released the Guide to Corporate Ecosystem Valuation (CEV), which provides information on how to quantitatively, or in some cases monetarily, assess risks and opportunities related to ecosystem services. CEV can therefore be a logical next step after undertaking an ecosystems services review.

8.2 IEMA self-test questions

Question 1

a) Outline the purpose of Environmental Impact Assessment (EIA) and Strategic Environmental Assessment (SEA). [6]
b) Outline the benefits that an environmental impact assessment can bring to a project. [6]

Question 2

a) Outline the main objectives of an environmental audit. [6]
b) Identify three advantages and three disadvantages of using an internal (first party) auditor and an external (third party) auditor. [6]

Question 3

a) Explain what is meant by 'ecological footprinting' and 'carbon footprinting'. [6]
b) Outline 3 advantages and 3 disadvantages of ecological footprinting. [6]

8.3 NEBOSH self-test questions

Question 1

Many organisations are interested in the integration of environmental, quality and health and safety management systems. Explain what you consider to be the advantages and disadvantages of such an approach. [20]

Question 2

BS EN ISO 14001:2004 requires organisations to set objectives and targets for assessing environmental management performance. With reference to appropriate examples:

a) **Explain** the difference between an objective and a target. [8]
b) **Outline** the factors that an organisation should consider in setting objectives and targets as part of its ISO14001 commitments. [6]
c) **Describe** an appropriate procedure for actions to be taken in the event that a target is not achieved by its due date. [6]

Question 3

a) **Identify** how a large scale electronics manufacturing company may contribute to the problem of climate change. [14]
b) **Explain** how the technique of life cycle analysis may be used to inform the organisation and help it reduce the extent to which its products contribute to climate change. [6]

Question 4

Prepare a presentation to your senior management team that **explains** the possible benefits to their organisation of implementing an Environmental Management System and gaining accreditation to ISO14001. [20]

8.4 Further information

Topic area	Further information sources	Web links (if relevant)
Aspects identification and environmental risk assessment	ISO14004:2004 – Environmental management systems – general guidelines on principles, systems and support techniques IEMA practitioner guide, 2006. Risk management for the environmental practitioner. IEMA	www.iso.org www.iema.net/shop
Environmental auditing	ISO19011:2011 – Guidelines for auditing management systems	www.iso.org
Environmental management systems	ISO14001:2004 – Environmental management systems – requirements with guidance for use EMAS:2009 – EC Eco-management and audit scheme	www.iso.org www.ec.europa.eu/environment
Resource efficiency surveys	WRAP and carbon trust case studies and assessment tools	www.wrap.org.uk www.carbontrust.com
Environmental footprinting	The living planet report and footprint calculators available from WWF	www.wwf.panda.org/
Life cycle assessment and eco-design	ISO14040:2006 Environmental management – Life cycle assessment – principles and framework ISO14006:2011 – Environmental management systems – guidelines for incorporating ecodesign	www.iso.org
Sustainable procurement	DEFRA publication, 2006. Procuring the future. DEFRA BS 8903:2010 – principles and framework for procuring sustainably – guide Environment Agency, 2003 – commodity sustainability briefings document	www.gov.uk www.bsigroup.com www.environment-agency.gov.uk
Carbon management	Carbon trust guidance and tools DEFRA, 2013 Environmental reporting guidelines – including mandatory	www.carbontrust.com www.gov.uk/defra

Topic area	Further information sources	Web links (if relevant)
	greenhouse gas reporting guidance. DEFRA	
Corporate ecosystems review	Hanson et al, 2012. The corporate ecosystems review – guidelines for identifying business risks and opportunities arising from ecosystem change. WRI publication. WBCSD publication 2011. Guide to corporate ecosystem valuation. WBCSD	www.wri.org www.wbcsd.org

CHAPTER 9

Analysing problems and opportunities to deliver sustainable solutions

This chapter is organised into the following sections:

9.1 Revision notes – impact assessment methods

9.1.1 Environmental modelling

Term describes a range of techniques used in EIA and pollution control to predict the physical, ecological and social outcomes of an activity.

Physical modelling – based on observations using a representative model, e.g. hydrological impacts of structures in a river or estuary

Mathematical (computer) modelling – based on computer models simulating key criteria. Typical applications are emissions plume models (to predict dispersion characteristics from chimneys), acoustic models (to predict noise transfer and attenuation) and traffic flow models (to predict the impact of increased vehicle movements at various times of day).

9.1.2 Statutory contaminated land assessment

The 2012 UK statutory guidance provides four distinct **grounds for designation as 'contaminated land'** as follows:

1. Significant harm is being caused to a human, or relevant non-human, receptor.
2. There is a significant possibility of significant harm being caused to a human, or relevant non-human, receptor.
3. Significant pollution of controlled waters is being caused.
4. There is a significant possibility of significant pollution of controlled waters being caused.

There is a risk based assessment method used to determine whether a site falls into one of these designations. This risk based approach is also applied to establishing remediation requirements, whether as part of statutory notices or voluntary agreements, aiming broadly:

a. to remove identified significant contaminant linkages, or permanently to disrupt them to ensure they are no longer significant and that risks are reduced to an acceptable level; and/or
b. to take reasonable measures to remedy harm or pollution that has been caused by a significant contaminant linkage.

9.1.3 Noise nuisance prediction using BS4142

BS4142 provides a standard methodology for measuring noise levels and predicting the likelihood of related nuisance, based on a comparison with background noise levels. There are 3 essential stages.

In practice these stages equate to the following steps:

Step 1 Measure or estimate the specific noise
Step 2 Adjust to standard time intervals if appropriate
Step 3 Adjustment 1 – based on reference to average background levels (LAeq without the specific noise)

Step 4 Adjustment 2 – based on annoyance factors (this gives us the 'rating noise level')
Step 5 Determine the background noise level (LA_{90})
Step 6 Subtract the background noise level from the rating noise level
Step 7 Compare the final number with the nuisance likelihood table to determine whether complaints are unlikely, of marginal significant or likely

BS4142 is used for risk assessment and decision making only not as a defence or a definitive guide as to nuisance control standards.

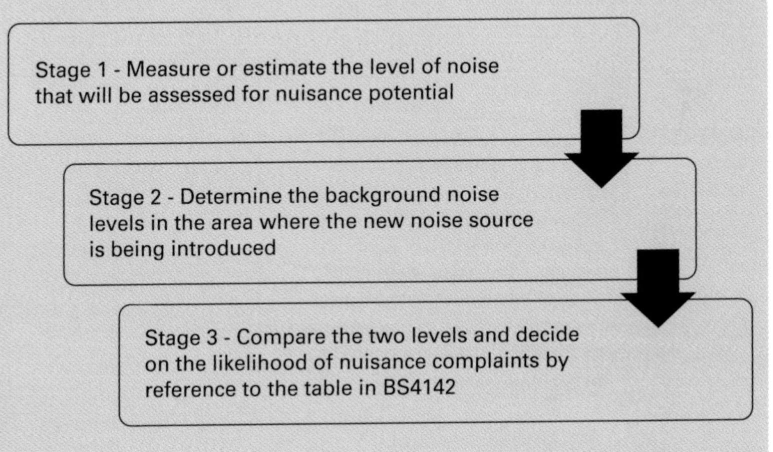

Figure 9.1 The three basic stages of BS4142

9.1.4 Planning guidance relating to noise nuisance

'National Planning Practice Guidance – Noise' is used by planning authorities to guide planning decisions that may create noise nuisance. Essentially the guidance works by classifying the effect of noise levels and applying a hierarchy of response from the perspective of the planning department.

Table 9.1 Planning guidance and noise nuisance

Perception	Examples of Outcomes	Increasing Effect Level	Action
Not noticeable	No effect	No observed effect	No specific measures required
Noticeable and not intrusive	Noise can be heard, but does not cause any change in behaviour or attitude. Can slightly affect the acoustic character of the area but not such that there is a perceived change in the quality of life.	No observed adverse effect	No specific measures required
		Lowest observed adverse effect level	
Noticeable and intrusive	Noise can be heard and causes small changes in behaviour and/or attitude, e.g. turning up volume of television; speaking more loudly; closing windows for some of the time because of the noise. Potential for non-awakening sleep disturbance. Affects the acoustic character of the area such that there is a perceived change in the quality of life.	Observed adverse effect	Mitigate and reduce to a minimum
		Significant observed adverse effect level	
Noticeable and disruptive	The noise causes a material change in behaviour and/or attitude, e.g. having to keep windows closed most of the time, avoiding certain activities during periods of intrusion. Potential for sleep disturbance resulting in difficulty in getting to sleep, premature awakening and difficulty in getting back to sleep. Quality of life diminished due to change in acoustic character of the area.	Significant observed adverse effect	Avoid

Perception	Examples of Outcomes	Increasing Effect Level	Action
Noticeable and very disruptive	Extensive and regular changes in behaviour and/or an inability to mitigate effect of noise leading to psychological stress or physiological effects, e.g. regular sleep deprivation/ awakening; loss of appetite, significant, medically definable harm, e.g. auditory and non-auditory	Unacceptable adverse effect	Prevent

Source: From the National Planning Practice Guidance – Noise

9.1.5 Predicting dust and odour nuisance

Risk factors that would need to be considered to determine nuisance potential include:

- the nature of the dust or odour (chemical constituents, health and property harm potential etc.)
- any relevant ambient air limits, e.g. PM_{10} set under the UK air quality strategy
- the sensitivity of receptors/neighbourhood characteristics, e.g. residential location or heavy industry area
- dilution and dispersion rates and conversely deposition rates and concentration points
- the presence of other sources of dust and odour in the same vicinity
- the frequency and scale of the dust/odour generation.

For **odour nuisance** DEFRA guidance suggests the use of the so called FIDOL factors:

Table 9.2 FIDOL factors determining the offensiveness of a particular odour

The FIDOL factors determining offensiveness	Comments
Frequency (How often an individual is exposed to odour)	Even an odour generally perceived to be quite pleasant can be perceived as a nuisance if exposure is frequent. At low concentrations a rapidly fluctuating odour is more noticeable than a steady background odour, i.e. this is an aggravating factor.

The FIDOL factors determining offensiveness	Comments
Intensity (the perceived strength of the odour, proportional to $\log_{10}$ concentration)	Generally the higher the intensity the more likely the nuisance.
Duration (the length of a particular odour event or episode, i.e. the duration of exposure to the odour)	Generally the longer the exposure the more likely the nuisance although see the comments about fluctuating odour above.
Offensiveness (based on a mixture of odour character and offensiveness/ attractiveness at a given concentration/intensity)	Some odours are universally considered offensive, such as decaying animal matter. Others may be offensive only to those suffering unwanted exposure in residential proximity, e.g. coffee roasting odour.
Location (the type of land use and nature of human activities in the vicinity of an odour source, i.e. the tolerance and expectation of the receptor)	Relates to the characteristics of the neighbourhood and the sensitivity of the receptors.

Source: After the DEFRA *Odour Guidance for Local Authorities*

For **dust nuisance** the Beaman and Kingsbury method uses comparisons of sticky pad readings of deposited dust with a reference table based on evaluation of case studies where nuisance has occurred.

Table 9.3 Dust level readings linked to probability of nuisance complaints

% effective area coverage/day	Predicted response
0.2	Noticeable
0.5	Possible complaints
0.7	Objectionable
2.0	Probable complaints
5.0	Serious complaints

Source: Based on categories from Beaman and Kingsbury, 1981

9.1.6 Ecotoxicity assessments and the use of indicator species

Biological indicator species (bio-indicators) are selected to be representative of pollution impacts to their ecosystem. Different indicator species may be required depending on the primary pollution pathway, as well as the specific ecosystem or pollutant characteristics.

Assessments may be done with acute concentrations of pollutants used to assess the lethal dose (LD) on the test organisms. Often expressed as LD_{50} – the concentration at which 50% of the target population will die. This is only really useful in cases of acute pollution, e.g. emergency planning.

More useful is an approach that attempts to estimate the *Predicted Environmental Concentration (PEC)* and the *Predicted No Effect Concentration (PNEC)* of pollutant releases into a specific receiving environment. PEC calculations typically involve modelling of pollutant release and dispersal. PNEC involves consideration of both lethal dose data and ecological monitoring data. An example of this approach is incorporated into the UK air quality strategy which uses ambient air limit values for key pollutants that have been set with reference to human health impacts.

9.2 Revision notes – control strategies

9.2.1 A general overview of pollution control techniques and emergency planning

Environment Agency/SEPA produce a series of Pollution Prevention Guidance (PPG) Notes covering a variety of pollution prevention and response issues.

Air pollution control examples – filtration systems, particulate separator systems, acid gas scrubbers, water curtains.

Water pollution control – examples of pollution prevention techniques – hydrocarbon interceptors, drain marking, closed loop vehicle wash systems, isolation mechanisms for drains during bulk deliveries, secondary containment for bulk storage, vandalism prevention measures, sealed drainage systems in flood plain areas.

Water pollution control – examples of effluent treatment techniques – settlement/flotation tanks, pH correction, cooling towers or lagoons, reed beds.

Land pollution control examples – dig and dump, barriers, soil washing, bioremediation.

Noise pollution control examples – management controls, e.g. operating hours, traffic management plans; technical controls, e.g. acoustic cladding, dampers.

Emergency planning – an important pollution prevention and control element.

9.2.2 Materials and waste management to reduce environmental impact

Important because of resource consumption and pollution associated with supply chain, manufacturing, use and disposal. Environmental impacts associated with materials usage may be reduced via three things:

- use less materials for the same level of activity, e.g. using returnable packaging
- use equivalent materials from suppliers that for some reason (e.g. location, operational standards etc.) have less environmental impact
- replace existing materials with high pollution potential with new materials with low pollution potential, e.g. replacement of solvent based with water based inks.

Impacts associated with waste disposal may be reduced by moving up the waste hierarchy, i.e. finding alternatives to landfill of waste in the following order of preference – waste prevention, reduction, reuse, recycling, recovery (waste to energy or composting).

9.2.3 Carbon management strategies focusing on building design and transport planning

Improving consumption efficiency in lighting, heating and power – using less overall energy per unit of work completed. Includes consideration of building design using reference standards such as BREEAM.

Increasing energy efficiency in transport:

- reducing transport (web based conferencing, route planning, supplier selection, employee car share schemes, home working, bike to work facilities etc.)
- reduced impact transport (efficient driving techniques, efficient vehicle selection including use of hybrid or electric vehicles, use of lower impact transport types, e.g. rail rather than road, sea rather than air).

Changing energy supply – to reduce associated pollution and increase the efficiency of supply. For example, gas as a primary fuel for electricity and space heating creates less impact than coal or oil, both through lower emissions output and reduced losses during energy or heat transmission. Better still on site combined heat and power plants may represent significant efficiency improvements. And of course the use of renewable energy – either self-generated or purchased as green tariff electricity – reduces impact still further.

9.2.4 Air pollution control techniques for particulates and gaseous pollutants

Three basic approaches:

a) **End of pipe technology** – including production and emissions control technologies.
b) **Substitution** – the substitution of less damaging substances for those with higher pollution potential, e.g. using water based rather than solvent based inks, low sulphur coal etc.
c) **Source controls** – involving the limitation (and even elimination) of polluting activities, e.g. car transport, energy usage and

thereby reducing associated emissions – may also include things like solvent pump dispensers rather than open cans to reduce fugitive emissions.

Selecting end of pipe control techniques

A number of factors must be determined before a proper choice of collection equipment can be made. Among the most important data required are the following:

- the physical and chemical properties of the pollutants;
- the range of the volumetric flow rate of the emission stream;
- the range of expected particulate concentrations (dust loadings);
- the temperature and pressure of the emission stream;
- the humidity and the nature of the gas phase (e.g. whether corrosive or not); and
- the required condition of the final emission.

End of pipe techniques – gaseous emissions

- Adsorption – 'adhering to' – techniques using a capture substrate such as activated carbon to which pollutants, e.g. solvents stick and are hence removed from the emissions stream. Pollutants may be recovered or the substrate filters disposed of as solid waste.
- Absorption – 'incorporation into' – scrubber systems use water vapour to absorb soluble emissions such as acid gases. An effluent is generated as a by-product but is relatively easy to treat.
- Condensation – conversion of gaseous pollutants into liquids using temperature or pressure control. Useful for recovery of high value volatile materials that can often be subsequently recycled.
- Combustion – thermal or catalytic conversion of pollutants into base constituents such as carbon dioxide, nitrogen and water. Used in vehicle exhausts – platinum catalysts help convert NOx into nitrogen and water.

- Bio-filtration – normally used in conjunction with adsorption techniques, e.g. carbon filter captures an organic solvent and bacterial cultures in the filter break down the pollutant into carbon dioxide and water. Requires steady stream of pollutant and control of temperature and humidity within the tolerance ranges of the organisms used.

End of pipe techniques – particulate emissions

- Gravity settling chambers – the emissions equivalent of a settling pond – particles settle out of air stream and in some cases can be classified by size and hence recyclability, e.g. shot blast systems.
- Cyclone separators – spinning the emissions stream creates forces which facilitate the settling out of particulates, e.g. use extensively for wood dust recovery in wood workshops/ manufacturing.
- Wet collectors – cyclone systems enhanced by the introduction of a water spray – increases particle recovery efficiency even further (although can create difficulty in terms of disposal of effluent/slurry).
- Electrostatic precipitators – use electricity to charge particles and then attract them from the air stream, e.g. used on a huge scale in steel works to capture iron ore and coal dust
- Fabric filters – used on a variety of scales from small scale local exhaust ventilation to three storey curtain systems in bakery environments. Fabric traps particles and allows air to flow through. Filters need changing regularly or required emptying systems.

End of pipe techniques – chimney design

The final stage in any end of pipe control system requires careful consideration of:

- The chimney height – as a minimum should be 0.6 times as high as surrounding buildings.
- Efflux velocity – minimum of 15m/s.

- Temperature of gases – relevant to ambient temperatures to enhance dispersal.
- The speed and direction of the prevailing wind – dispersal modelling in relation to key receptors.
- The concentration of a pollutant – higher concentrations require greater dispersal characteristics in chimney design.
- Atmospheric conditions, e.g. temperature inversions can completely undermine the dispersal of pollutants so care in the siting and design of chimneys where such phenomena occur is vital.

9.2.5 Effluent treatment techniques in both public main sewer and industrial pre-treatment plants

Trade effluent may be discharged with or without pre-treatment to sewer or controlled waters. Treatment techniques are considered as main sewage treatment works and as industrial pre-treatment processes.

Sewerage treatment process

- Biological oxidation – aerobic bacteria
- Nitrogen removal – aerobic and anoxic bacteria
- Phosphorous removal – aerobic bacteria and chemical precipitation
- Sludge treatment and disposal (including peat/soil replacement, fuel for power generation, anaerobic digestion).

Industrial pre-treatment processes

Physical treatments

- Simple settlement or flotation tanks to remove organic or inorganic solids (often used in conjunction with chemical flocculants or precipitants which aid in separation of the solids).

- Adsorption using carbon filters to concentrate and remove impurities.
- Centrifuge and cyclones to separate out solid contaminants.
- Oil/water separators (flotation or tilted plate types) to remove oil and grease.
- Reverse osmosis (also known as ultra-filtration) – a membrane technology used to separate out dissolved solids such as salts (this method may also be used to regenerate process water for reuse).
- Solvent extraction – evaporation tanks with vapour recovery units.

Chemical treatments

- Ion exchange or manipulation to remove metal impurities or convert them into less harmful ions (e.g. conversion of hexavalent chrome to trivalent chrome reduces its toxicity significantly).
- pH regulation via the addition of acid (dilute hydrochloric acid is commonly used) or alkali (dilute sodium hydroxide is often used).

Biological treatments

Biological treatment is perhaps best established in the form of reed beds – artificial wetlands which often have additional wildlife benefits. Useful in particular for hydrocarbon and suspended sediment removal.

9.2.6 Contaminated land remediation techniques

Removal – excavation of contaminated material and disposal at a licensed waste facility.

Containment – involving on-site engineering to contain the parcel of contamination using methods such as surface capping, vertical and/or horizontal barriers to prevent migration of contaminants.

Treatment – this may occur on or off site and may include physical, biological or chemical methods to reduce or eliminate the hazards associated with contamination. Examples include – air stripping/sparging to facilitate bioremediation or the recovery of volatile materials; groundwater pumping into wells or boreholes to facilitate recovery of floating contaminants such as hydrocarbons; in situ bioremediation of soil and/or groundwater pollution; soil washing; thermal and chemical treatments in mobile or static facilities.

Project CL:AIRE is a public/private partnership that provides valuable independent assessment and case study reporting of new and established remediation techniques.

9.2.7 Solid waste disposal options

As the waste hierarchy suggests prevention is the best control strategy for waste. All other options involve some kind of environmental impact as shown in Table 9.4.

Table 9.4 Waste management options – advantages and disadvantages (author)

Waste management option	Advantages	Disadvantages/impacts
Landfill	Is relatively low cost Possible to generate electricity from captured landfill gas Design and construction techniques are tried and tested	Non-productive use of land Source of noise, odour and vermin nuisance Generates hazardous leachate which may contaminate ground water Production of methane gas No secondary value obtained from resources
Incineration	Reduces volume of waste significantly – leaving only residual ash May destroy otherwise very persistent hazardous materials	Generates air emissions including some high hazard pollutants such as dioxins and lead

Waste management option	Advantages	Disadvantages/impacts
	Possible to generate (low carbon intensity) electricity as an integral process	High capital investment in plant provides little incentive for waste minimisation initiatives Second time use of resource only
Anaerobic digestion	Possible to generate (low carbon intensity) electricity as an integral process (but without the emission issues associated with incineration) Reduces volume of waste significantly – leaving only residual sludge	Second time use of resource only
Composting	Produces a useable product that may also reduce the impact associated with peat based alternatives Reduces the biodegradable material sent to landfill and hence reduces landfill gas generation	Odour nuisance Variable quality input can significantly affect product quality
Recycling	Multiple resource re-use possibilities which may mean retained value for waste producer Avoidance of landfill Avoidance of further 'new materials' resource consumption	Pollution and resource consumption associated with the recycling process Technical difficulties, lack of infrastructure and market limitations in relation to recycled products
Reuse	As for recycling	Transport (and perhaps storage) impacts in getting the materials to point of reuse

9.2.8 Control approaches for noise, odour and dust nuisance

Noise (reference – EA H3 horizontal guidance note)

Engineering controls – noise barriers, insulation, screens, low noise equipment, damping mats, jacking piling rather than percussion techniques, background sensitive reversing alarms.

Lay out controls – location and orientation of buildings/noise generating equipment.

Administrative – operating hours, traffic routes, maintenance schedules, operating procedures, stakeholder consultation regarding noisy operations.

Odour (reference – EA H4 horizontal guidance note)

- Adsorption using activated carbon, zeolite, alumina
- Dry chemical scrubbing using solid phase systems impregnated with chlorine dioxide or potassium permanganate
- Biological treatment including soil bed bio-filters, non-soil bio-filters – peat/heather, wood-bark, compost – bio-scrubbers
- Absorption (scrubbing) using spray and packed towers, plate absorbers
- Incineration using existing or dedicated boiler plant, thermal or catalytic systems
- Other techniques including odour modifying agents, condensation, plasma technology, catalytic iron filters and ozone and UV.

Dust (reference – London code of practice)

Haul roads/site entrances – temporary surfacing, road sweeping, damping down, speed limits (especially on unpaved surfaces), load covers, wheel washing.

Excavations/earthworks – damping down, covering or re-vegetation.

Stockpiles/storage mounds – minimise drop heights, wind screens, storage time limitations etc.

Waste management – prohibition on burning, storage times, covers etc.

9.2.9 Sustainable procurement strategies

Supplier specification – i.e. dictating standards relating to company operations or product design and/or materials content.

Partnership approach – i.e. working with suppliers to collectively find solutions to unsustainable or high impact products or processes.

Selection approach – using available information to choose between suppliers and/or products such that the sustainability impact of our choice is reduced.

The choice of which strategy or strategies an organisation employs will depend on its level of commitment, its purchasing power, its relationship with its first tier suppliers and the amount of information available in relation to the environmental impact of products and services purchased.

9.2.10 Biodiversity protection and enhancement strategies

Biodiversity benefits associated with 'traditional' environmental management:

- Reducing the scale of waste, pollution and nuisance generated by an organisation which may help reduce direct pressure on individuals, species or habitats.
- Reducing the quantities of raw materials and utilities consumed by the organisation which may have implications in terms of pressure on biodiversity throughout the supply chain.

More focused initiatives may also be undertaken under the following headings:

Direct initiatives

- Land management for conservation in line with local biodiversity action plans

- Recognition for conservation efforts via the Wildlife Trust's biodiversity benchmarking scheme.

Indirect initiatives

- Voluntary support of local conservation initiatives via funding or volunteer time
- Supply chain initiatives through the use of selection criteria in purchasing, e.g. use of forestry stewardship scheme accredited timber or paper
- Biodiversity offsetting – using DEFRA guidance to ensure that any habitat destruction associated with a development is 'compensated' by the creation of equivalent or greater areas of equivalent or higher value habitat.

9.3 IEMA self-test questions

Question 1

a) Outline three environmental risks that may arise from the bulk storage of chemicals on site. [3]
b) Propose a model for assessing environmental risk. [3]
c) Apply the model proposed in b) to two of the risks identified in a) noting any assumptions made. [6]

Question 2

The development of a holiday park comprising a number of self-catered lodges, a swimming pool, catering and retail units is proposed for an area of woodland close to a small rural town.

Identify six environmental impacts that could arise during construction of the part and six that could arise during operation. For each impact, propose a possible mitigation measure. [12]

Environmental impact – construction	Mitigation measure

Environmental impact – operation	Mitigation measure

Question 3

a) Explain what is meant by 'environmental risk'. [4]

b) Describe four consequences for an organisation of failing to manage environmental risk. [8]

9.4 NEBOSH self-test questions

Question 1

Environmental modelling is widely used in assessing the risks arising from releases of a pollutant into the environment. **Describe**, using appropriate examples, the issues that might be considered in a model assessing air pollution from stack emissions at a planned industrial site. [20]

Question 2

A heavy manufacturing company situated close to residential property is considering the introduction of new work patterns involving evening and weekend work.

a) Making reference to appropriate standards, **describe** how the potential for environmental noise nuisance should be assessed. [12]

b) **Outline** the types of control methods that might be considered should a noise problem be shown to be likely to occur. [8]

Question 3

A chilled foods manufacturer is in the process of inviting tenders for a contract for the servicing and maintenance of refrigeration systems at its site. Amongst other requirements, the contract will include the removal of all related wastes including electrical equipment, refrigerant gases and oils.

Describe the *environmental* issues the organisation should consider *when selecting a contractor*. [20]

9.5 Further information

Topic area	Further information sources	Web links (if relevant)
Assessment techniques	DEFRA publication – Contaminated land statutory guidance 2012 BS4142 Method for rating industrial noise affecting mixed residential and industrial areas National Planning Practice Guidance – Noise	www.gov.uk/defra www.bsigroup.com http://planningguidance.planningportal.gov.uk/
Carbon management strategies	BREEAM guidance on sustainable building design	www.breeam.org
Contaminated land remediation techniques	Project Claire guidance and case studies	www.claire.co.uk/
Noise Control techniques	Environment Agency H3 Guidance note	www.environment-agency.gov.uk
Odour control techniques	Environment Agency H4 Guidance note	www.environment-agency.gov.uk
Dust control techniques	Mayor of London best practice guide – the control of dust and emissions from construction and demolition	
Biodiversity strategies	World business council for sustainable development – business ecosystems training tool Wildlife Trust biodiversity benchmark DEFRA biodiversity offsetting guidance and pilot project information	www.wbcsd.org www.wildlifetrusts.org/biodiversitybenchmark www.gov.uk/biodiversity-offsetting

Developing and implementing programmes to deliver environmental performance improvement

This chapter is organised into the following sections:

10.1 Revision notes

10.1.1 Implementing an environmental improvement programme

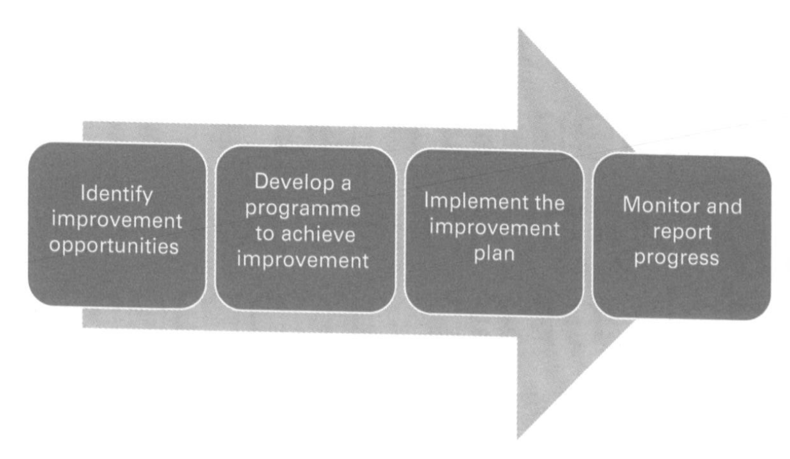

Figure 10.1 Steps in the implementation of an environmental improvement programme

Identifying improvement opportunities

Using resource efficiency surveys, life cycle assessment, environmental auditing etc.

Develop a programme

Comprising policies, objectives and SMART targets backed up by action plans containing tasks, timescales and responsibilities.

SMART targets:

- **S**pecific
- **M**easurable

- **A**chievable
- **R**esponsibilities allocated
- **T**ime bound

Implement the improvement plan

Including tracking and reviewing progress using audits etc. to ensure we stay on track and record the completion and impact of individual actions.

Develop methods to monitor and communicate progress

Using key performance indicators (see Chapter 7) and effective internal and external communication techniques (see Chapter 11).

10.1.2 Creating a business case for change

General drivers for change (see also Chapter 6) under the categories of:

- Legal
- Financial
- Market
- Social

Pitching a specific change programme often involves consideration of cost implications, sales implications and indirect benefits. In such a context we should consider the following.

Present financial information in clear cost benefit analysis terms

Often easy to do (see example in Table 10.1) with resource efficiency initiatives although care should be taken to state any assumptions made. Not so easy to quantify savings associated with risk reduction measures but attempts can be made using clean up cost estimates etc.

Table 10.1 A simple example of a cost benefit analysis (author)

Cost of new equipment installation	£2,000
Cost of disposal of old equipment	£500
Savings in running costs per year associated with new equipment (at current pricing levels)	£500
Return on investment timescale	5 years
Estimated total net project lifetime savings (based on the predicted life of the new equipment – in this case 20 years)	£7,500

Present ethical benefits in relation to key stakeholders

Clearly identify key stakeholders and the benefits arising from the improvement proposal from their perspective.

Consider non-financial drivers (including clarity over environmental benefit)

Non-financial drivers may cover a range of things from legislative compliance through to demonstration of environmental benefits. For example, if government policy and consultation around legislation suggests that a particular course of action may be required in the mid to long term, a case could be made that improvements made now will be easier to manage and cause less disruption than waiting until mandatory requirements appear. Conversely, clarity over actual environmental benefits, e.g. the creation of new habitat as part of a wildlife focused landscaping scheme, makes the non-financial benefits clear as opposed to the rather woolly statements about 'environmentally friendly' developments or investments.

Benchmark performance standards and consider trends

This means consciously positioning ourselves in relation to competitors and in relation to change trends. Some organisations may choose to position themselves as best in class and leaders in environmental innovation. Some aim for mid field performance, while still others may choose to do the minimum required. Each might be an appropriate strategy for an organisation – the key is to ensure that positioning in terms of environmental performance and credibility matches positioning in terms of customer base and considers business sector pressures.

Task

Identify an issue in your organisation where you consider there to be a real opportunity for improvement. It can be an improvement in relation to pollution control or resource efficiency or perhaps even both. Prepare an outline business case for 'selling' your idea to senior management in search of support and/or funding. You should organise your case under the following headings:

- The proposal
- The cost benefit analysis (use estimates if necessary)
- The ethical benefits and their relevance to key stakeholders
- Non-financial drivers/benefits.

10.2 IEMA self-test questions

Remember that information from other chapters may be useful in answering any question. Although themed in relation to the IEMA syllabus elements candidates are expected to draw on relevant information from any part of the syllabus and/or their own experience in answering exam questions. This is notably the case in some of the example questions below.

Question 1

You have been asked to help your local school to reduce its water consumption.

a) Describe the approaches you would use. [6]
b) Describe how you would monitor the effectiveness of the approaches you propose. [6]

Question 2

You have been asked to run an energy awareness campaign for your organisation's administrative staff.

a) Outline the key messages you would include in your campaign. [6]
b) Describe how you would monitor the effectiveness of the campaign. [6]

Question 3

a) Identify approaches that can be used to improve environmental performance. [4]
b) Describe how you would identify the specific environmental training required in an organisation. [8]

10.3 NEBOSH self-test questions

Question 1

a) **Outline** the key elements of the so-called 'waste hierarchy'. [8]
b) Using suitable examples, **describe** how the 'waste hierarchy' could be implemented in relation to office waste. [12]

10.4 Further information

Topic area	Further information sources	Web links (if relevant)
Improvement plans and environmental management systems	ISO14004:2004 EMS General guidelines on principles, systems and supporting techniques	
Creating a business case	The Times 100 business case studies – especially the business and the environment section WRAP resource efficiency case studies	businesscasestudies.co.uk/case-studies www.wrap.org.uk

Communicating effectively with internal and external stakeholders

This chapter is organised into the following sections:

11.1 Revision notes

11.1.1 Identifying environmental stakeholders and their interests

Table 11.1 Stakeholder groups and example areas of interest (author)

Stakeholder	Example interests
Government	Wants industry to respond to government policy and measures, including the uptake of voluntary initiatives such as ISO 14001 or EMAS.
Regulators	Want companies with regulated activities to demonstrate compliance and the capability of continually delivering compliance.
Customers	Business customers may be exerting pressure for suppliers to demonstrate environmental responsibility and improve environmental performance. Some may even be taking life cycle environmental considerations into account with consequences for procurement decisions. Competitors may be developing and demonstrating effective environmental management thus increasing the pressure.
Consumers	The so-called 'green pound' is a notoriously difficult thing to quantify, however, some market sectors have emerged in recent years that depend on consumer interest in environmental issues. Furthermore, judging by purchasing patterns associated with high profile incidents, consumers increasingly will not buy from companies that have a poor environmental record or have products associated with high profile environmental problems.
Financial institutions	The developing environmental agenda and environmental concerns are presenting new risks for financial institutions as providers of capital or insurance. For example, investors may require more information on potential environmental liabilities, capital expenditure associated with environmental problems and the effect of environmental issues on profits. Public reporting of environmental performance in annual accounts is becoming an issue.
Public	How companies manage their environmental issues and demonstrate the effectiveness of this management influences the public image of the company – possibly affecting sales, complaints (especially from the local community), and in some cases even their license to operate. Green groups and the

Stakeholder	Example interests
	media can have a key role in influencing this aspect of a company's reputation (positive or negative).
Parent companies or corporate functions	Subsidiaries or individual sites of multi-site organisations are often subject to expectations from parent companies or group functions to demonstrate compliance with a set of performance standards aimed at providing assurance in relation to corporate responsibility or reputation management/brand positioning.
Directors	Company directors not only have a business interest in pollution control and resource efficiency, but also a personal interest in legal compliance. If a company is shown to be poorly managed or negligent in its control of environmental issues, Directors can be held personally liable and prosecuted directly under the UK Environmental Protection Act, 1990.
Employees	Employees are unlikely to be keen to work for employers with poor environmental performance or a bad image. For some employees at least, good environmental practices and a sense of personal involvement in implementing improvements will act as a workplace motivator.
Non-governmental organisations (NGOs)	Lobby groups and local interest groups come in a wide variety of sizes, geographical spread and specific interests. At each scale, however, they may have a considerable influence on an organisation's operations and as such should certainly be considered key stakeholders.

Task

Consider your own organisation and compile a list of key environmental stakeholders and their areas of interest.

Stakeholder	Area of interest

Table 11.2 NGOs as protectors of the environment – the pros and cons (author)

Advantages	Disadvantages
• credibility in terms of public opinion • autonomy from the electoral process • international perspective of both problems and potential responses • greater flexibility in confronting polluters or unacceptable behaviour • promote 'grass roots' action and personal engagement in environmental issues • strong media profiles ensure widespread coverage of actions or campaigns	• insecurity of funding • limited access to information • dependence on fund raising and/or volunteer support • difficulty in cooperation/coordination with other NGOs or government bodies • distrust/barriers between NGO and target organisation • high profile stunts may alienate parts of the target audience

11.1.2 Effective stakeholder communications

As with any area of communication there are a number of issues to be considered in environmental programmes to ensure that clear understanding and engagement is achieved.

Internal and external stakeholder needs

Closely tied to the identification of stakeholders as above but should also be applied to different audience groups within the organisation. Consideration of needs should include:

- what might they need/want to know or say
- how do they prefer to receive information/provide feedback.

ISO14001 sets specific requirements for communication with internal and external stakeholders including communication of roles, reporting of performance, policies, controls and priorities etc.

Communication objectives

Four key headings describing **generic motives** for communication:

- Informing
- Instructing

- Motivating
- Consulting

Under these headings there may be a wide number of environment specific objectives including:

- To ensure rules are followed
- To encourage involvement in environmental improvement initiatives
- To minimise risk due to inappropriate action
- To identify opportunities for improvement
- To improve community relations
- To promote company profile in the public arena
- To raise awareness and credibility with customers and/or consumers
- To ensure good relations with regulators
- To comply with statutory requirements.

What to communicate

The actual content of any communication depends on the objective but there are some general principles that prove good 'rules of thumb':

- Simplify but do not dumb down – almost everyone prefers concise and accessible information but that does not necessarily mean trivialising the subject.
- Make the message relevant to the audience – generally people are most receptive to information relating to their experience and to issues that they have the capability to influence. How it relates to them and what they can do about them may also be part of the message.
- Avoid overload – it is better to stick to fewer strong messages that get through than swamp the audience and risk none getting through.
- Make it interesting – it is sometimes necessary to be imaginative, innovative, artistic and even downright dramatic to achieve this.

Table 11.3 Internal communications – examples of methods (author)

Method	Strengths	Weaknesses
Training and workshops	Provide good opportunity for interaction, clarification and confirmation of understanding.	Can be highly dependent on delivery and ineffective if insufficiently focused and engaging. Can be expensive and time consuming.
Posters	Low cost and low disruption. Can be highly effective if simple, clear, engaging and appropriately positioned.	No opportunity for feedback or clarification. May be 'unseen' if inappropriately positioned or not changed frequently.
Video/e-Learning	Can provide a reasonable level of interaction and confirmation of understanding. May be more flexible and hence less disruptive/costly than 'face to face' training.	Can be difficult to tailor the message to a particular audience/ work environment.
Meetings/ briefings	Good opportunity for two way communication and immediate feedback of issues, concerns or clarification requirements.	Time consuming and can be difficult to maintain consistency of communication over a larger workforce if a 'cascade communication system' is employed.
Brochures/ newsletters	If done well can be an efficient way of communicating with a large group of people especially if geographically spread out. Graphics and photographs can make for visually interesting information.	No opportunity for feedback or clarification. May be 'unseen' if poorly produced or too frequently used.

Table 11.4 External communications – examples of methods (author)

Method	Strengths	Weaknesses
Brochures and newsletters	If done well can be an efficient way of communicating with a large group of people especially if geographically spread out. Graphics and photographs can make for visually interesting information. Web based/electronic versions can be very cost effective.	No opportunity for feedback or clarification. Can be dismissed as propaganda by critical stakeholders if not presented in a credible or even verified format.
Community liaison committees	Can be very effective in ensuring that stakeholders feel heard and included in decision making. Affords opportunities to receive information that may be useful to the organisation.	Can be time consuming and difficult to manage. Can be difficult to engage with representative individuals – community communication is generally reliant on feedback from committee members.
Helplines/ complaints procedures	Provides an easy way for those with issue to be heard and feel like they are being responded to. Can prevent problems escalating.	Are essentially reactive in nature, i.e. the problem has already occurred.
Regulator consultation and reporting	Clear, timely and accurate routine reporting and incident or issue based liaison can help develop a sense of confidence in the regulator that an organisation is in control and 'doing the right thing'.	Can be time consuming but as a legal requirement this can hardly be seen as a weakness. May occasionally be made challenging when individual personalities clash.
Green claims/ marketing	If standards of reporting are credible and effective (see DEFRA guidance) then market share benefits may be gained through clear differentiation of product or company.	May backfire if claims are not clear and possible to substantiate. May be very time consuming and costly to complete at a high standard of assurance and credibility.

Method	Strengths	Weaknesses
Corporate credibility schemes, e.g. ISO14001	Simple, third party assured method of providing basic credibility at least between organisations.	Poorly understood/ recognised by consumers. Do not necessarily differentiate between basic competence and high level performance.
Product credibility assessment	Schemes such as the EU eco-label scheme, the Forestry Stewardship Council mark etc. provide third party assurance of standards linked to specific stakeholder concerns.	Limited coverage of such third party schemes means that it can require sector consultation and support to develop credible standards.

11.1.3 Corporate environmental reporting

Denmark, New Zealand and the Netherlands all have legislation on environmental reporting. From April 2013 in the UK all 1800 FTSE listed companies are legally required to report on greenhouse gas emissions. Following a review in 2015, this may be extended to all large companies (i.e. those with over 250 employees). This mandatory greenhouse gas reporting is being introduced because of the clear belief that the requirement to report on performance provides a strong motivation to improve.

Reporting goals vary but might include one or more of the following:

- Educate stakeholders
- Explain/promote the organisation's achievements
- Reassure regulators that further legislation is not required
- Persuade stakeholders of the 'green credentials' of the organisation
- Counter negative press or claims by environmental groups
- Legitimate the industry or product sector
- Express personal commitment by the board
- Signal to financial stakeholders that environmental risk is managed sensibly.

There are a number of reporting standards that are being increasingly recognised and used by organisations to ensure that reporting is credible. Key examples include:

- DEFRA corporate reporting guidance
- The European Eco management and audit scheme (EMAS)
- The global reporting initiative
- ISO14063:2006 Environmental management – Environmental communication – Guidelines and examples
- AA1000AS:2008 – data assurance standard for monitoring and reporting.

The Global Reporting Initiative

The goals of this approach are as follows:

1. define a standard sustainability reporting format,
2. encourage voluntary third party assurance that the reporting standards have been achieved, and
3. collate reports in a searchable database to enable benchmarking of corporate reports and sustainability performance.

The GRI protocols are aimed at standardising data collation and ensuring transparency regarding the definition of reporting boundaries. The guidance requires a series of 'standard disclosures' under the following headings:

Strategy and profile: such as company strategy, description, key impacts, risks and opportunities, key stakeholders and governance issues such as commitments to widely recognised standards such as ISO14001 or the sustainability principles set out in Agenda 21 from the Earth Summit in Rio.

Management approach: disclosures that cover how an organisation addresses the 'material aspects' (i.e. those relevant to them) outlined in the strategy and profile section. This is intended to provide context for understanding performance in specific areas.

Performance indicators: indicators that elicit comparable information on the economic, environmental and social performance of the organisation. Core indicators are specified in each category but with guidance relating to the tailoring of the data to make it organisation specific.

The whole approach is voluntary but provides a credibility benchmark for organisations involved in corporate reporting.

Corporate social responsibility reporting

Corporate social responsibility (sometimes referred to as corporate governance) is even less well defined than sustainability but may be thought of as encompassing the responsibilities that an organisation has to all its stakeholders.

It is perhaps simplest to consider short term CSR strategies as being a 'moral business approach', while long term CSR strategies are indistinguishable from a 'sustainable business approach'.

11.2 IEMA self-test questions

Remember that information from other chapters may be useful in answering any question – although themed in relation to the IEMA syllabus elements candidates are expected to draw on relevant information from any part of the syllabus and/or their own experience in answering exam questions. This is notably the case in some of the example questions below.

Question 1

a) Identify six stakeholders that an organisation may have and, for each stakeholder, identify the environmental information in which they may have an interest. [6]

b) Identify six benefits that an organisation may gain from effective communication with its stakeholders. [6]

Question 2

a) Identify five stakeholder groups that may be interested in the environmental performance of an organisation and identify the types of environmental information they may require. [5]

b) Describe the purpose and main elements of the Global Reporting Initiative. [7]

Question 3

a) Explain what is meant by a 'green claim' and explain why they are important. [4]

b) Green claims should be:

i. Clear
ii. Accurate
iii. Relevant
iv. Substantiated

For each of these characteristics identify two elements of good practice when making a claim. [8]

11.3 NEBOSH self-test questions

Question 1

Identify four reasons why an organisation may choose to make public an annual report on their environmental performance and outline the typical content that might be included within such a report. [20]

11.4 Further information

Topic area	Further information sources	Web links (if relevant)
Green claims	DEFRA, 2011. Green claims guidance. DEFRA	www.gov.uk/defra
Corporate reporting	DEFRA, 2013 Environmental reporting guidelines – including mandatory greenhouse gas reporting guidance. DEFRA The Global Reporting Initiative, 2013. G4 – sustainability reporting guidelines. Reporting principles and standard disclosures. GRI Brady et al., 2011. Environmental Management in Organisations – the IEMA Handbook, 2nd ed. Earthscan	www.gov.uk/defra www.globalreporting.org

CHAPTER 12

Influencing behaviour and implementing change to improve sustainability

This chapter is organised into the following sections:

12.1 Revision notes

12.1.1 Understanding culture change

'Culture change' may be defined as a fundamental shift in attitudes and behaviours in an organisation that eventually becomes self-sustaining and independent of management systems and/or individual champions.

The following characteristics might be important aspects of an organisation's 'culture':

- Dominant styles of communication – e.g. consultative vs prescriptive.
- Dominant thinking styles – e.g. creative and visual vs analytical and data based.
- Decision making styles – e.g. inclusive and local vs top down/mandate based.
- Experience of change – e.g. successful and supported vs imposed and less than fully successful.
- Company – employee relationship – this multi-faceted area encompasses the balance between employee and management perceptions. It might incorporate issues such as employees' sense of security and individual worth, as well as the company belief in a motivated and innovative workforce.
- Scale and uniformity of the organisation – disparate and distinct areas of operation vs a common identity and shared goals.

In relation to environmental management and sustainability, culture change might mean the development of a company ethos that could be anything along a spectrum ranging from 'prevent on-site pollution' to 'become a completely sustainable company'.

12.1.2 Obstacles to culture change

Some of the most commonly heard reasons for not adopting environmental improvement initiatives include:

- Not enough time
- Too expensive
- No benefit to the organisation

- Customer constraints
- Changes too risky – what if we get it wrong?
- Not enough expertise
- People don't care
- No commitment from senior management.

Less easily heard but perhaps often present in individual reactions to change programmes might be the following:

- This sounds like more work for me with no more reward.
- What's wrong with the way we currently do it?
- This won't work because . . . why won't they listen to me?
- This sounds like it will get in the way of me doing my job – my supervisor will not like it.
- I don't understand what benefit we'll get from doing it this way.

Depending on circumstances any of these reasons could be objective or subjective and might be held by one individual or by the majority of the workforce.

Task

Identify key obstacles to implementing environmental improvements in your organisation. Record your key obstacles in the table below:

	Subjective	**Objective**
Individual	Based on one person's world view, assumptions or attitudes.	Based on one person's role, authority, skills, resources etc.
Collective	Group culture, shared mind sets	Political, economic, social, technological, legal, environmental conditions

12.1.3 Constructively dealing with resistance

> Resistance is a natural response and instead of thinking about 'overcoming obstacles', think about 'supporting employees' through the change.
>
> (The Prosci Change Management Learning Center)

Key principles of support might include:

What does it mean for me? – often initial resistance is based on a concern around this question. Immediate clarity on this can avoid a hardening of resistance or misinterpretation.

Hearing it from the right person – people respond much better when communications explaining the change objective and implications for them come from either a senior management figure or their direct supervisor (and ideally both).

Covert resistance – organisations vary in terms of their communication openness – it is often best to assume that there will be some covert resistance to change. Look for it so that it can be brought into the open and addressed.

An example

For example, if we are trying to introduce a waste segregation system in a production area, we may, through discussion with people in the area, discover that there is considerable scepticism around municipal recycling schemes and concerns that the requirement to segregate waste may make already time pressured tasks even more difficult to perform. We might therefore predict resistance around two key objections which could be summarised as 'there's no real benefit – it will all get landfilled anyway' and 'I'm not going to have enough time to do this'.

The first 'subjective' obstacle will need to be addressed both initially and in terms of ongoing communication, with transparency and explanation (especially if arrangements are such that occasional loads do have to be landfilled). The second objection (which may or may not be objective) will need to be carefully considered in putting in place the arrangements for segregation. The best way to address this area of resistance would probably be to involve the operators themselves in deciding how the segregation process could work best.

Communication relating to the justification of the segregation process and assurances around clarity of disposal arrangements may best come from senior management or team leaders. The environmental team may then get involved alongside team leaders to help agree the segregation arrangements with the teams.

12.1.4 Planning for change

Obviously a lot of overlap with continual improvement loop seen in environmental management systems but emphasising consultation and communication. In essence:

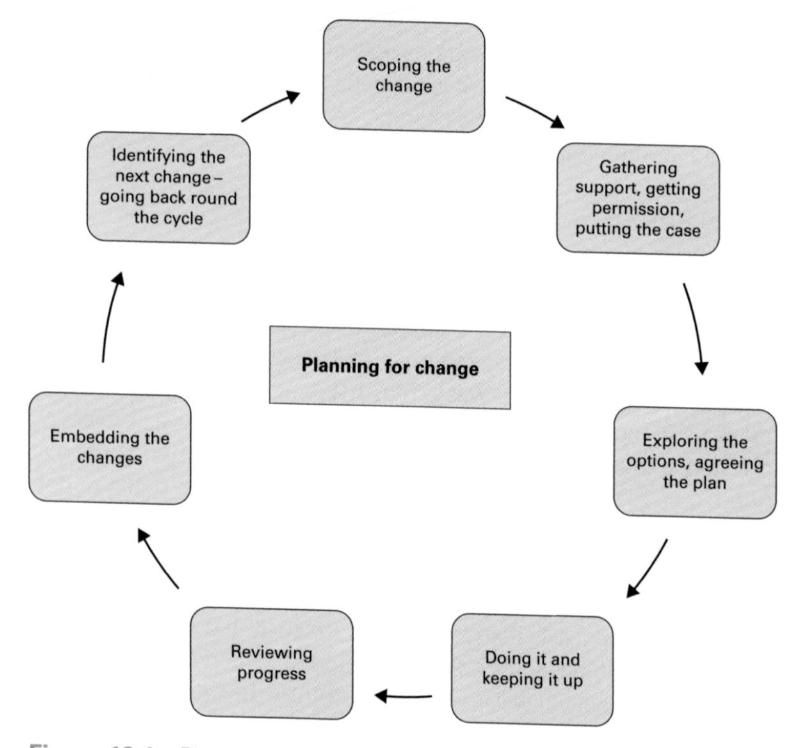

Figure 12.1 Planning for change

- Be clear about what we want to change/achieve and why.
- Talk and listen to people a lot and in equal amounts.
- As a change leader – walk the talk but be realistic about how far and how fast progress can be made.

12.2 IEMA self-test questions

Remember that information from other chapters may be useful in answering any question – although themed in relation to the IEMA syllabus elements candidates are expected to draw on relevant information from any part of the syllabus and/or their own experience in answering exam questions. This is notably the case in some of the example questions below.

Question 1

a) Outline the principles of change management. [6]
b) Outline the factors which may prevent an organisation from adopting a more sustainable approach to its management. [6]

Question 2

a) Outline six benefits that an organisation may derive from a decision to take a more proactive approach to environmental management. [6]
b) Outline the process of change management that would support such a move. [6]

Question 3

You have been asked to implement an environmental strategy for your organisation.

a) Outline the possible barriers to implementation. [6]
b) Outline the arguments/approaches that you would use to overcome these barriers. [6]

12.3 NEBOSH self-test questions

Question 1

Prepare a memorandum to an organisation's senior management explaining the advantages and disadvantages of a systems style approach using the ISO14001 model as a way of controlling and improving its environmental performance. [20]

Question 2

Many organisations include environmental awareness in their company induction training for new employees.

a) **Outline** the environmental content that should be included in such a programme. [10]

b) **Describe**, with appropriate examples, ways in which all employees might be encouraged to engage with a company's environmental management programme, with particular reference to procedure compliance. [10]

12.4 Further information

Topic area	Further information sources	Web links
Change management	The mindtools website has a host of free and subscription management information, a lot of which relates to change management.	www.mindtools.com
The psychology of culture change	There is a lot about change management on this website but of particular interest is the section on psychological contract theory.	www.businessballs.com/changemanagement
Change management	An excellent tutorial series that covers much of the content of this chapter in much more depth and with links to further information and research.	www.change-management.com/tutorial-7-principles-mod1

APPENDIX 1

IEMA specimen answers

The specimen answers presented below are not approved or issued by the IEMA. They are based on the experience of the author over many years in delivery of training programmes, review of examiners feedback and the sample marking schemes provided by the IEMA.

Chapter 3 Question 1

a) *Describe either the carbon cycle or the nitrogen cycle. [4 marks]*
b) *With reference to two contrasting examples, describe how human interventions impact upon natural cycles. [8 marks]*

(a) The carbon cycle describes the storage and movement of carbon between the geological, biological, atmospheric and oceanic reservoirs.

Geological carbon is stored in the earth's crust in two principle forms – carbonaceous rocks such as limestone and chalk, and fossil fuels. Storage is long term and stable.

Biological carbon is stored in plant and animal tissues and storage is relatively short term by comparison to the geological reservoir. Carbon is released to the atmosphere in the form of carbon dioxide during tissue decomposition following death.

Atmospheric carbon is in the form of carbon dioxide and is relatively short term storage as it is absorbed into the oceans and also fixed into biological carbon by plants during photosynthesis.

Oceanic carbon is in two forms – dissolved carbon dioxide and the carbonaceous shells of marine organisms. Dissolved carbon dioxide may be released into the atmosphere and the carbonaceous shells may in the long term enter the geological reservoir in the form of new sedimentary rocks.

(b) The combustion of fossil fuels on a widespread and global basis primarily for transport and energy generation has a major impact on the carbon cycle. Carbon is released from storage as geological hydrocarbons in the most long term reservoir into the atmosphere in the form of carbon dioxide. This represents a huge escalation of the normal transfer rate from the geological to the atmospheric reservoirs and has resulted in an unprecedented increase in global atmospheric carbon dioxide concentrations. This trend has been and continues to be further compounded by the large scale clearance of forest areas globally. Such clearance and the associated burning of plant material releases huge quantities of biological carbon previously held in plant tissues into the atmosphere, again in the form of carbon dioxide.

The knock on consequence in the carbon cycle of the escalating concentrations of atmospheric carbon dioxide has been the increased concentration of dissolved oceanic carbon dioxide which forms carbonic acid and leads to a gradual but perceptible acidification of the earth's oceans. Of course the major consequence of these human disturbances to the carbon cycle is climate change.

A contrasting example of human intervention may also be seen in relation to the hydrological cycle. Urbanisation on a large scale leads to a big increase in paved surface area. This impacts the rate of run-off into rivers and streams. A high rainfall event in an area of low paving would run off into surface waters but also be absorbed into soils where it may slowly percolate to surface waters or may be taken up by plant roots and transpired back to the atmosphere. Once paved all the rainfall is channelled via drainage systems direct to surface waters leading to a sometimes dramatic increase in flood levels. In this way the human impact on the natural cycle is an acceleration of the transfer rate between atmospheric water storage and surface water systems.

Chapter 4 Question 2

a) *Identify four advantages and four disadvantages of using fiscal instruments as an environmental policy instrument. [4]*

b) *Describe two fiscal instruments currently in use that deal with different environmental problems, stating the country in which they apply. [8]*

a)

Advantages of fiscal instruments	Disadvantages of fiscal instruments
Provide a commercial incentive to behave in a particular way	If set too low they may not lead to behaviour change
May be used to generate public funds to further support the policy in question	If not 'ring fenced' they may be perceived as general taxation rather than linked to an environmental goal
Provides flexibility of response in those targeted by the instrument	May disadvantage commercial organisations competing in markets where no such fiscal instruments apply
May involve less administration than permitting or other forms of statutory controls	May be expensive to administer and enforce depending on the collection systems applied

(b) The UK climate change levy is a tax paid by all commercial users of energy for lighting, heating and power. Such energy use is a major contributor to UK national greenhouse gas emissions and hence climate change. The levy applies to multiple energy types including gas, electricity and fuel oil and is collected from the user at point of sale by the energy providers. The providers then pass on the tax to HM revenue and customs. The tax is ring fenced for spending on projects supporting the UK climate change programme targets. The purpose of the climate change levy is to provide a clear incentive to organisations to become more energy efficient (and hence reduce tax payments as well as energy purchase costs). It also encourages investment in renewables and/or high quality combined heat and power which are both exempt from the levy.

The UK landfill tax is a tax paid at point of landfill of commercial and industrial wastes. The tax is collected by landfill site operators. A grant scheme exists to allocate some of the money collected to community sustainability projects. However, the majority of the money collected goes to central government. Ultimately, waste producers pay the tax as the waste carrier contractors include it as part of their fees. The result is an incentive for waste producers to segregate waste in order to find alternative disposal routes to landfill. For example reuse, recycling and waste to energy. There are two bands of tax applied – a low level inert waste fee and an active waste fee that has increased annually since the tax was introduced in 1996. Figures suggest that the tax has been highly successful in decreasing the disposal of commercial and industrial waste to landfill and opening up new markets in recycling and material reuse.

Chapter 5 Question 1

You are the environmental manager for a company that produces 800kg of hazardous waste per year.

a) *Identify the key legislation for the control of hazardous waste. [2]*
b) *Identify the regulator responsible for enforcement of this legislation. [1]*
c) *Describe the main actions that the company must take to comply with this legislation. [9]*

a) The Hazardous Waste (England and Wales) Regulations 2005, Environmental Protection Act Part II section 34 (the duty of care), Waste (England and Wales) Regulations, 2011.
b) The Environment Agency (England) and Natural Resources Wales (Wales).
c) The Hazardous Waste Regulations 2005 essentially require an organisation to identify and segregate hazardous waste (as defined in the WM2 guidance note) from other types of waste. As the organisation produces more than 500kg of hazardous waste per year it must also register as a hazardous waste producer with the Environment Agency. Registration is done on line and must be renewed annually. Registered companies are issued with a 'premises code' which must then be included on all waste transfer notes. A summary of all transfers made must be kept and may be subject to review by the regulator.

Compliance with the Environmental Protection Act Part II section 34 (the duty of care) requires the company to store the waste appropriately in order to prevent its escape, only transfer it to an authorised waste carrier, and keep accurate and complete waste transfer notes for three years. As part of the latter accurate classification of the waste must be made in accordance with the European Waste Catalogue codes found in the List of Waste (England) Regulations, 2005.

Compliance with the Waste (England and Wales) Regulations, 2011 requires that the organisation apply the waste hierarchy to the waste it generates including the hazardous waste and to make a statement on each transfer note that it has endeavoured to minimise and/or find alternatives to disposal in order to reduce the quantities of waste sent to landfill.

Chapter 6 Question 2

Outline the business benefits of taking a proactive approach to environmental management. [12]

1. Reduced likelihood of legal non-compliance and hence risk of prosecution and fines/legal costs
2. Reduced insurance costs where risk control is part of premium review
3. Improved relationships with regulators potentially leading to reduced permitting fees and inspection schedules
4. Reduced waste disposal costs through increased recycling and waste minimisation programmes
5. Reduced raw material costs through resource efficiency measures
6. Reduced energy costs through energy efficiency measures
7. Increased sales opportunities through company or product reputation
8. Improved mid to long term planning in relation to energy and raw materials supply security
9. Improved relations with neighbours and local lobby groups
10. Improved reputation with shareholders and ethical investment schemes
11. Improved employee relations with recruitment and retention benefits

12. Improved access to finance where lenders apply environmental risk management criteria.

Chapter 7 Question 1

You have been asked to establish a system for monitoring your organisation's waste production and disposal. Describe the data you would collect and explain how you would present this data in order to monitor performance against targets over time. [12]

Essentially we need to obtain data relating to the amount of waste generated, where it is generated and how it is dealt with. Data useful from these perspectives would include:

Input materials data including raw materials, consumables and any associated packaging. This could most usefully be compiled in the form of data normalised to levels of company activity. For example kilogrammes of material X used per kilogramme of product produced – compile on a daily, weekly, monthly basis depending on how easily available the information is. Once an understanding of usage rates and opportunities for improvement exists, targets could be set and progress tracked on a monthly basis using charts showing actual ratio vs target ratio per month.

Waste generation data could similarly be tracked and if possible broken down into different waste streams and even by departments or operational units. Again, while total figures may be compiled and reported in absolute terms, it is likely to be more useful to normalise the data to activity levels. So for example kilogrammes of packaging waste per 1000 units packed in the despatch area. Again this allows target setting and performance indicators that apply regardless of variable activity levels so that if, for example, twice as many units are packed in month 2 the same indicator and target could be used as in month 1.

Disposal route data may also be useful in tracking the environmental impact of the waste generated. A very typical example is to monitor the percentage of waste sent to landfill. The implication here is that less waste sent to landfill constitutes an environmental performance improvement. Targets can then relate to a reduction in percentage which again allow for continuity of monitoring regardless of variable activity levels by the organisation. Presentation of the data graphically, with actual vs target percentage lines, gives a good visual representation of performance.

In all cases data can most usefully be presented in highly visual formats, e.g. graphs and charts, ideally at or close to the activities most influential in waste generation or management. This helps maintain awareness of the waste management targets and progress to date and improve employee motivation to act appropriately.

Chapter 8 Question 2

a) *Outline the main objectives of an environmental audit. [6]*
b) *Identify three advantages and three disadvantages of using an internal (first party) auditor and an external (third party) auditor. [6]*

a) Environmental audits may vary widely in their goals and objectives, as there are many very different investigations grouped under the general term environmental audit. Broadly speaking we can think of two main categories of environmental audit – assurance audits and status audits.

Assurance audits include internal environmental management systems (EMS) audits which aim to evaluate compliance with an organisation's policies and systems. External EMS audits aim to evaluate compliance of an organisation's management system with an external standard such as ISO14001. A legal compliance audit will consider the degree to which an organisation meets legal requirements in general or perhaps will consider a specific legal obligation, e.g. waste management duty of care auditing. Assurance audits are often repeated periodically as part of a company assurance process.

Status audits tend to be completed as a research exercise either to facilitate a decision (e.g. due diligence auditing which aims to identify liabilities as part of a company take over or land purchase) or to inform and commence an improvement programme. Waste reviews, energy audits, material usage evaluations fall into this latter category and they focus less on policies and procedures and more on recording current practices and activities relevant to the topic area. The intention with these kinds of status audits is to generate findings that are opportunities for improvement of some kind related to the topic under consideration. These findings will often then become the basis of a subsequent management programme perhaps linked to environmental targets.

b)

	Internal auditor	External auditor
Advantages	Familiarity with company activities and procedures can improve efficiency of audit	High level of third party assurance credibility associated with findings of external auditors
	Employees may be more willing to open up to a colleague.	May bring expertise of good practice and/or standards from other sites/companies
	Typically lower cost that external resource	Findings may be taken more seriously by internal decision makers
Disadvantages	May be less likely to challenge traditional practices	May be inefficient in the audit due to a lack of understanding of process or organisation
	May have personal involvement in some areas audited which could lead to a lack of objectivity	Auditees may be reluctant to be open and helpful with external auditors as it may be perceived as 'betraying' the organisation
	May have limited expertise or experience particularly in areas such as legal compliance	Typically much higher cost that internal audit resource

Chapter 9 Question 2

The development of a holiday park comprising a number of self-catered lodges, a swimming pool, catering and retail units is proposed for an area of woodland close to a small rural town.

Identify six environmental impacts that could arise during construction of the part and six that could arise during operation. For each impact, propose a possible mitigation measure. [12]

Environmental impact – construction	**Mitigation measure**
Noise nuisance from plant and equipment	Planned working hours to minimise disturbance

Traffic nuisance arising from staff and materials movement to and from site	Traffic management plans to minimise disturbance, use of crew buses etc.
Habitat loss associated with site clearance	Pre-construction surveys to ensure no protected species loss. Habitat creation as part of landscaping activities.
Silty water run-off from excavations and exposed sub-soil	Construction of temporary settlement lagoons, use of 'silt busters' or similar.
Spillage of fuels or construction chemicals causing land contamination	Use of drip trays and bunding wherever fuels and chemicals are stored or used.
Waste disposal impacts at landfill sites used for construction waste	Waste management plans to minimise waste generated and maximise segregation for recycling.
Environmental impact – operation	**Mitigation measure**
Noise nuisance from residents of the holiday park	Installation of acoustic fencing and/or use of site lay out to minimise disturbance
Traffic nuisance to neighbours from access to park	Speed controls, road improvements and signage to avoid areas where nuisance potential is greatest
Climate change associated with energy use on park	High standard of building insulation and on site solar power generation to supplement grid bought electricity
Waste disposal impacts at landfill sites used for operational waste	Waste management plans to minimise waste generated and maximise segregation for recycling.
Flooding caused by accelerated run-off from hard standing areas of park	Installation of sustainable urban drainage systems that use lagoons to slow feed rain water off site into receiving water bodies
Wildlife disturbance in remaining woodland on site	Provision of bat boxes, bird boxes etc. to encourage wildlife in remaining habitat.

Chapter 10 Question 2

You have been asked to run an energy awareness campaign for your organisation's administrative staff.

a) *Outline the key messages you would include in your campaign. [6]*
b) *Describe how you would monitor the effectiveness of the campaign. [6]*

a)

Stakeholder group	Environmental information requirements
Facilities management personnel	Energy usage breakdown and any opportunities for improving system settings through use of building management systems for space heating, reducing temperature setting on hot water systems etc.
All staff	Switch off campaign for all personal equipment when not in use, e.g. computers, printers
Users of printers and photocopiers	Messages focused on double sided printing and/or 'do you really need a copy/print' – ideally linked to consumption data, e.g. a photocopier left on overnight uses enough electricity to boil so many cups of tea etc.
All staff	Lights off message ideally linked to quantified costs and impacts of lighting
All staff	Space heating and cooling messages – put on a jumper to allow less heating; open the window to allow less air conditioning etc.
Cleaners/security staff	Clear procedures for switching off specified equipment left on after the end of the working day
All staff	Campaign targets and regular updates of key performance indicators and progress trends.

b) There might be a number of ways to usefully monitor the effectiveness of such a campaign. Tracking key performance indicators (including total energy use, electricity usage, space heating fuel usage etc.) to look at progress against targets is the highest level method, and if broken down, may give an indication of where improvements have been most effective (perhaps reflecting the greatest success of the awareness messages). However, more 'ground floor' monitoring might include:

- Staff interviews with random personnel to assess familiarity with campaign targets, progress and personal implications.
- Switch off surveys of lights or equipment at lunch times or evenings to look at the degree of compliance with switch off awareness messages.

- Photocopier and/or printer usage data could be monitored by area to assess local improvements in reducing usage.
- Auditing of individuals or groups with key responsibilities, e.g. the switch off checks conducted by evening workers, the degree day monitoring and heating adjustments by facilities management staff.

Chapter 11 Question 2

a) *Identify five stakeholder groups that may be interested in the environmental performance of an organisation and identify the types of environmental information they may require. [5]*

b) *Describe the purpose and main elements of the Global Reporting Initiative. [7]*

a)

Stakeholder group	**Area of interest**
Regulator	Compliance data required as routine reporting conditions set by permit. Any variation to permitted operations.
Shareholders	Legal compliance status and reputation reporting including corporate reports, benchmarking in sector, ISO14001 certification etc.
Residential neighbours	Pro-active community projects, prior notice/ consultation regarding site development or changes to operations that may have local nuisance or pollution implications.
Employees	Company reputation and standards – compliance status with third party standards such as ISO14001. Internal progress information regarding to environmental targets. Also role specific information related to expected actions or precautions required of the individual or employee group.
Customers	Company reputation and standards – compliance status with third party standards such as ISO14001, plus legal compliance status, product related information, e.g. hazardous material content etc.

b) The purpose of the Global Reporting Initiative is to provide a consistent and credible standard of environmental performance reporting. It has three key areas of focus:

1. Definition of a standard sustainability reporting format,
2. Encouraging voluntary third party assurance auditing to demonstrate that the reporting standards have been achieved, and
3. Collating reports in a searchable database to enable benchmarking of corporate reports and sustainability performance.

The standard format sets out generic approach and requires organisations to include generic context information related to their areas of activity, corporate policies etc. It then requires organisations to report against all relevant key performance indicators. Third party assurance of report contents and source data can be done to different degrees of detail. Companies are encouraged to adopt the level of assurance appropriate to them and then to declare the degree to which they have complied with the GRI reporting and assurance standard.

The collation of reports and standardised guidelines for reporting are an important attempt to drive up standards of transparency and accountability by organisations for their environmental/ sustainability standards. As an assurance mechanism this approach provides readers of corporate reports, whether customers or potential investors, with confidence that what they are seeing is a reliable and complete assessment of a company's sustainability status.

From an organisation's perspective the expected benefit is a clarity of the required coverage of reports so that they do not need to 'reinvent the wheel' but have instead a recognised format to follow.

Chapter 12 Question 1

a) *Outline the principles of change management. [6]*

b) *Outline the factors which may prevent an organisation from adopting a more sustainable approach to its management. [6]*

a) Change management is the systematic process of planning, implementing and embedding a change of practice and often associated values within an organisation. Key principles are as follows:

- Scoping the change – researching the issue, identifying priorities and building a business case around what is required and what benefits are expected to the business, the environment and key stakeholders.

- Gathering support – this requires the identification of key figures who will be needed to support, manage or implement the change. The case must then be made on an individual or group basis, taking account of the prevalent values, interests and potential resistance that may be held by different parties. The aim at this stage is to gain support for the overview not to present a fait accompli detailed plan of how a change will be achieved.
- Exploring the options and agreeing a plan – this involves further consultation with those involved so that agreement is consensus based and the rate and nature of change is both realistic and in line with the change management objectives. A formal agreement of key actions, milestones and targets by all parties can help engender a sense of ownership, responsibility and personal involvement in the process.
- Doing it and keeping it up – essentially the process at this stage is monitoring and communicating the change actions being undertaken to ensure transparency of progress. Unforeseen resistance or obstacles may need to be dealt with and adjustments made as required. It is critical at this stage that key change leaders are seen to be 'walking the talk'.
- Reviewing progress – in addition to ongoing monitoring of actions and key performance indicators, periodic review of whether the change process as a whole is heading in the direction required, whether outcomes are proving as expected and whether any change of emphasis or approach is required.
- Embedding the changes – what happens at this stage is highly variable but it is essentially about making sure that the change achieved becomes a routine part of the way the organisation works. It may involve adaption of policies or procedures, training or contract terms but all of these formal mechanisms are essentially supporting the establishment of a new 'custom and practice' within the business.

b) Factors creating obstacles to sustainability programmes include:

- Lack of management commitment – leading to mixed messages and priorities being received by the workforce
- Lack of financial resources – blocking the investment in equipment, people or services required to achieve a change
- Lack of time – on the part of key personnel
- Lack of understanding of the business case – at best this can lead to sustainability being seen as a moral luxury, at worst an unwelcome diversion
- Resistance to change amongst employees – often focused around a lack of understanding of what is required, what the benefits might be and what the implications are for the individual
- Lack of relevant skills and knowledge – even when the goal is accepted this can mean people are unable to identify appropriate actions.

APPENDIX 2

NEBOSH specimen answers and marking schemes

The specimen answers and marking schemes presented below are not approved or issued by NEBOSH. They are based on the experience of the author over many years in delivery of training programmes, review of examiners reports and assistance of delegates through the assessment process.

Chapter 3 Question 2
Describe *the ways an organisation might attempt to minimise its local and global adverse impacts on plant and animal communities (biodiversity). [20]*

Answer plan

Describe = detailed factual account of key features

- Biodiversity – definitions and overview of concerns
- Measures to minimise direct impacts
  - On site habitat protection and/or creation including the identification and monitoring of biological indicators

  - Grounds maintenance activities – minimise use of pesticides, fertilisers etc.
  - Water run-off and abstraction designed to protect and enhance biodiversity (e.g. SUDs)
  - Links to local biodiversity action plans, projects and the identification of key receptors; perhaps working with Wildlife Trust assessors if appropriate
  - Control of releases in normal operations and also emergency plans that take account of biodiversity
  - Planning considerations when expanding, building or relocating
- Measures to minimise indirect impacts
  - LCA and/or supply chain assessment focusing on biodiversity impacts
  - Supplier selection criteria/questionnaires to raise awareness and generate information
  - Use of registration schemes such as Forestry Stewardship Council approved timber, organic agricultural products
  - Product design that considers elimination of materials with biodiversity risk – either from a resource consumption or a pollution potential

Full answer

Biodiversity is defined as the variety of life in a given area expressed in terms of habitat, species and genetic diversity. Appropriate natural diversity in healthy ecosystems is considered to be desirable not simply from a moral or aesthetic viewpoint but from a very humanistic view of the provision of what have become known as 'ecosystem services', e.g. the delivery of raw materials and assimilation of wastes from human activity.

Organisations may have a variety of direct and indirect impacts on plant and animal communities and hence influence biodiversity both locally and on a more regional or even global scale.

Measures to minimise direct impacts

Typically these measures relate to activities carried out by organisations on their operating site. Clearly the scope increases in proportion to the land managed by an organisation but highly proactive measures have been implemented even by organisations with minimal land holdings in urban areas (through the use of log piles, window boxes, roof gardens etc.).

Measures to identify, protect and even create useful wildlife habitat on an organisation's site can include the planting of native species, cultivation of wetland areas or the creation of ponds etc. While not necessarily visually unattractive the prime aim of such land management practices are wildlife focused rather than landscaping or aesthetic. Monitoring of populations of key species identified or considered desirable may allow organisations to contribute to Local Authority Biodiversity Action Plans or data collation for the regional countryside body. Working with Wildlife Trusts or the Local Authority can provide the organisation with a lot of information and advice as to the nature of local priorities and potential contributions that could be made on their site.

Even where habitat creation is less actively promoted the careful and minimal use of pesticides, fertilisers and even things like mowing regimes may have a significant benefit in protecting wildlife locally.

Surface water management on larger sites may also be a very important issue to maintain consistency of flows in nearby water courses – avoiding flood peaks and low flow periods through reduced groundwater recharge (both the latter being associated with large areas of hard standing and traditional culvert drainage systems). Sustainable urban drainage systems which aim to regulate run-off rates may also generate useful new wetland habitat.

A whole range of pollution control and emergency response measures may be appropriate for organisations with the potential to pollute habitat, soils or water courses. As well as minimising the potential for harm during normal operations, special measures may be mobilised during emergency scenarios to provide adequate protection for locally important habitat.

Finally consideration of habitat issues should be a part of an organisation's planning process when expanding, building or relocating – seeking to minimise habitat damage and maximise habitat creation as part of any development.

Measures to minimise indirect impacts

In addition to the direct measures described above, many organisations can have a significant influence on biodiversity on a wider scale. The first task is often to understand the biodiversity issues in our supply chains. Life cycle assessment for key products can be an invaluable tool in this respect – identifying the potential for habitat loss. However, full scale life cycle assessment is both costly and time consuming and there are many generalisations that an organisation can usefully employ. Sourcing wood and paper products from Forestry Stewardship Council accredited sources, for example, provides a degree of assurance that forests are being managed with consideration of wildlife issues. Similarly organic products or locally sourced materials both provide assurance of lower general environmental impacts and hence risk to plant and animal communities.

In a more general sense organisations can include questions about biodiversity protection in supplier questionnaires as a way of raising supply chain awareness of concerns and customer interest in this area.

Finally and very importantly, organisations can consider improving the design and delivery of their own products and services to reduce biodiversity risk in any of the direct or indirect areas mentioned previously, either through the reduction of pollution potential or through the specification of materials that have an inherently lower biodiversity risk, e.g. lower hydrocarbon content or avoidance of hazardous materials that may pose an end of life pollution risk etc.

Chapter 4 Question 1

a) **Summarise** *what is meant by the term* 'best available technique' *in the context of the Pollution Prevention and Control Act 1999. [5]*

b) ***Describe** the scope and key requirements that may be imposed by a permit to operate a part A1 'regulated activity' as listed in the Environmental Permitting (England & Wales) Regulations, 2010. [15]*

Marking scheme	Available marks
Part a)	
Statutory context – EU IPPC Directive, UK Pollution Prevention and Control Act 1999, Environmental Permitting (England and Wales) Regulations 2010	1
Best practice standards agreed at European level involving consultation with permit holders at trade association level – defined for the 'regulated activities' listed in the Environmental Permitting Regulations	1
Set out as operational guidance in the BREF notes	1
Coverage – technical specifications, equipment design, material usage, operating procedures, monitoring arrangements, supervision, training and competencies etc.	2
Used by regulator to determine permit applications and consent requirements – operators must meet BAT standard for their process	1
Part b)	
Description of regulator (EA) and permit application process – consideration of BAT as defined in BREF notes, and local operating environment	1
Permit conditions • Site condition report and post closure requirements • Groundwater monitoring and protection measures • Emission limit values for releases to air/land/water • Raw material usage • Waste management standards • Energy efficiency requirements • Emergency planning – prevention and response • Nuisance management especially noise, dust and odour • Staff competence and training • Management system requirements • Record keeping requirements	up to 12
Monitoring and reporting requirements (routine and abnormal)	2

• For emission streams (using defined UKAS/MCERT standards) • For operational controls, abnormal emissions and incidents Regulator inspections • Based on perceived risk using the OPRA scheme	1
Improvement programmes • Actions to be completed in accordance with defined timescales – binding in the same way as other conditions	1
Payment for regulatory input – linked to OPRA assessment – application and annual subsistence fees	1

Chapter 5 Question 3

A chemical manufacturing organisation operates a Part A(1) process under the Environmental Permitting (England & Wales) Regulations under a permit issued by the Environment Agency.

a) *Outline the enforcement options open to the regulator in the case of non-compliance with the conditions of the permit. [15]*

b) *Explain how the organisation would proceed and what they would need to demonstrate to the regulator if they wish to surrender their permit. [5]*

Answer plan

Part a)

- Intro – EA is regulator for A1 processes
- Enforcement notice – breach of permit but continued operations allowed
- Suspension notice – breach or pollution threat – no continued operations
- Revocation notice – 'cessation and remediation' notice
- Prosecution potential and penalties
- Appeals procedure available to operator

Part b)

- Application to regulator
- Decommissioning arrangements
- Site survey and if necessary remediation plan

Full answer – part a)

The Environmental Permitting Regulations 2010 require all 'regulated activities' to operate within the terms of a permit issued by the appropriate regulator. A1 processes are regulated by the Environment Agency (EA) under powers granted by the Pollution Prevention and Control Act 1999. In addition to being able to set operational standards to be met by the permit holder at the time of granting approval, the EA has powers to issue three distinct types of statutory notice under different circumstances.

Failure to meet the requirements and conditions set within any of these statutory notices constitutes a criminal act and may lead to prosecution. Penalties include fines of up to £50,000 and/or up to 12 months imprisonment on summary conviction in a magistrate's court or unlimited fines and/or up to 5 years imprisonment on indictment in a crown court. The regulator may also seek to recover any clean-up costs associated with pollution offences.

The different enforcement options open to the regulator are as follows.

Enforcement Notice

This statutory notice may be issued in the event of:

- a breach or risk of breach of the permit conditions or where
- actual or potential pollution related to the regulated activity but not subject to permit conditions exists.

Action required will be defined in general or specific terms along with timescales for completion. The permit holder may continue to operate the regulated activity while the improvement actions are being carried out but beyond the notice date, operation of the process constitutes an offence unless the conditions of the enforcement notice have been complied with.

Suspension notice

This statutory notice may be raised where there is a more serious risk of actual or potential pollution arising from the regulated activity.

As with the enforcement notice, the breach of permit or risk of pollution will be identified along with the required remedial actions to be taken and the timescales for completion. However, when a suspension notice is raised no further operation of the permitted process is allowed until the remedial steps are completed and the notice has been withdrawn by the regulator. Statutory offences in relation to this kind of notice include failure to complete the required remedial actions and operating the process prior to the withdrawal of the notice by the regulator.

Revocation notice

This statutory notice may be issued when circumstances are such that the regulator wishes to withdraw the permit to operate the 'regulated activity'. This may be done in the event of permit compliance failure or as part of a planned site closure or cessation of operations. In either case the notice will specify actions to be completed to make the site safe and if necessary to clean it up to a satisfactory condition. As above it is a prosecutable offence to fail to comply with the terms of the notice within the defined timescales.

In all permits, an appeal process is available to the operator (via the Secretary of State) if they feel that the conditions attached to the permit are inappropriate, the timescales for completion too onerous or that the cause of the pollution or permit breach that precipitated the notice is not their responsibility.

Full answer – part b)

Surrender of a permit issued under the Environmental Permitting Regulations requires an application to be made to the regulator together with an appropriate fee. The application must set out the arrangements for closure of the site/process in a way that controls the risk of pollution and that returns the site to a satisfactory state. Reference to site conditions recorded during the pre-commencement site survey is assumed wherever such information is available.

The regulator may set additional decommissioning conditions, site survey requirements (including on-going monitoring) or remediation actions as appropriate as part of the approval process. Failure to complete any of these agreed permit surrender actions may lead to prosecution.

Chapter 6 Question 1
Describe *the business reasons for effectively managing environmental risk. [20]*

Answer plan

- Legal
- Financial
- Social
- Market

Full answer

There are an increasing number of reasons why environmental risk should be managed which may be broadly grouped under the following categories.

Legal pressures

The coverage of environmental law has widened significantly since the early 1990s with the scale of fines, damages and remediation costs also escalating. At its most basic, environmental risk should be managed so as to avoid prosecution and its associated negative financial and reputation implications. But beyond this, the management of environmental risk is now an inherent element of numerous statutes. Organisations working under a permit issued under the Environmental Permitting (England and Wales) Regulations 2010, or who are registered under the Control of Major Accident Hazard Regulations 2005, are required to have carried out risk assessments and to have implemented prevention and response measures that ensure that pollution is avoided or minimised. Similarly new statutory powers under the Environmental Damage Regulations enable regulators to take action against those responsible for pollution or the management of operations that create significant risk of pollution. In both these latter categories, the failure to effectively manage risk can itself lead to regulatory action – a pollution incident need not have occurred.

Financial pressures

These come in a variety of forms in relation to risk management. The costs of fines and punitive measures has been alluded to above but within this category should be added the costs of civil claims associated with property or personal damages. Growing awareness of these kinds of liabilities has also led to greater focus on due diligence and inherited liability. Failure to manage the risk of land contamination for example can lead to significant loss of value of premises or land holdings. The clean-up costs for such contamination is often also very high.

Lending and insurance companies are becoming increasingly aware of the liabilities transferred to them and are bringing increasing pressure on organisations to demonstrate high standards of risk management in order to secure loans or insurance cover.

Shareholders and corporate investors are increasingly looking at the environmental credentials of organisations with which they are associated. This is partly defensive, i.e. avoiding being associated with an organisation that may be subject to prosecution or negative press as a result of a failure to adequately manage environmental risk. But it is also becoming a pro-active strategy to make high standards of environmental risk management in both the short term and longer term an investment criteria. The justification being on ethical grounds or as an indicator of sound management more generally.

Finally, and perhaps of most immediate importance to most organisations, the operational cost savings associated with resource use efficiency, waste reduction and reductions in abatement/treatment costs can provide a strong reason to manage environmental risks effectively.

Social pressures and opportunities

The general rise in public awareness of environmental issues has led to increasing pressures from both internal and external stakeholder groups to manage environmental impacts effectively. Local residents are much less tolerant of nuisance or perceived pollution issues than previously. Poor reputations in local communities can lead to increased incidence of complaints as well as objections to planning or permit applications.

Employees are increasingly interested in the reputation of their employer. Companies with high environmental standards report implications in terms of staff morale, recruitment and retention.

Market pressures and opportunities

Increasing public and corporate awareness is also translating into market pressures and opportunities with selection for companies or products with good credentials and against those perceived to be polluters. Business customers increasingly have some kind of supplier evaluation process that is looking to discriminate against those organisations perceived to be failing to manage environmental risk. The consumer market has been similarly selective as companies involved in major pollution incidents have found to their cost, e.g. Shell and BP.

For all of the above reasons a systematic and comprehensive approach to environmental risk management has become a working requirement for many organisations.

Chapter 7 Question 2

a) *Oxides of nitrogen (NOx) are an important group of air pollutants.* ***Describe*** *the main environmental impacts arising from their emission into the atmosphere. [14]*

b) ***Outline*** *two methods that could be used to quantify NOx emissions generated by an organisation. [6]*

Marking scheme	Available marks
Part a)	
Sources and NOx variants – NO, N_2O, NO_2 – transport, combustion processes etc.	1
Ground level impacts to human and animal health – respiratory impacts – both directly (NO_2 in particular) and through the generation of secondary pollutants in photochemical smog	3
Harm to plants and ecosystems through toxic and nutrient addition impacts (local eutrophication)	3
Odour nuisance	1
Wet and dry deposition (acid precipitation) leading to damage to infrastructure and buildings	3
Acid precipitation impacts in sensitive (acid) ecosystems leading to terrestrial and aquatic impacts via corrosive, toxic and secondary pollutant (e.g. mobilisation of metal ions in soils) effects	3
Climate change contribution as N_2O is a greenhouse gas.	2
Part b)	
Ion chromatography – including outline description of technique	3
Chemiluminescence – including outline description of technique	3
Infrared spectrometry – including outline description of technique	3
Ultraviolet spectrometry – including outline description of technique	3

Chapter 8 Question 3

a) ***Identify*** *how a large scale electronics manufacturing company may contribute to the problem of climate change. [14]*

b) ***Explain*** *how the technique of life cycle analysis may be used to inform the organisation and help it reduce the extent to which its products contribute to climate change. [6]*

Answer plan

Part a)

- Explanation of the greenhouse effect – natural process, human influence and key GHGs

- Direct (site based) GHG emissions – processes, boilers, refrigerants, solvent usage, transport
- Indirect GHG emissions from energy, waste disposal, staff transport
- Indirect GHG emissions from supply chain – manufacturing and infrastructure and including carbon sink removal
- Direct (product based) GHG emissions – energy use, maintenance/repairs and end of life disposal

Part b)

- Define life cycle and LCA
- Describe product unit based calculations of GHGs
- Provides ability to compile a full life cycle GHG impact – embedded carbon at individual stages or overall
- Provides ability to compare the impact of design changes on carbon footprint

Full answer – part a)

The greenhouse effect is a natural process whereby gases in the earth's atmosphere trap reflect long wave length radiation from the sun as it bounces back from the earth's surface. It is the phenomenon responsible for the moderation of temperatures at the earth's surface. However, since the industrial revolution human activity has resulted in the emissions of large quantities of the gases associated with the greenhouse effect. The gases involved are varied and come from a variety of sources but many are combustion based and in terms of quantity carbon dioxide and methane are the two most significant.

A manufacturing plant may directly and indirectly contribute to the Greenhouse Effect in a number of ways that might be considered under the following headings.

Direct site based contributions

Manufacturing processes that emit greenhouse gases such as carbon dioxide, oxides of nitrogen, many types of solvent or methane contribute directly to the increasing concentration of these gases in the atmosphere and hence the intensification of the greenhouse effect.

Similarly gas or oil fired boilers used for space or water heating emit similar gases.

Air conditioning or refrigeration systems very often use gases that have global warming potential and any leaks or losses during top up or replacement of such systems will also contribute to the greenhouse effect.

Transport activities related to the delivery of goods to customers as well as work related travel by staff are further sources of greenhouse gas emissions for many manufacturing companies.

Indirect site based contributions

Activities that do not fall directly under the manufacturer's control but nonetheless are an integral part of their operations may be another source of contribution to the Greenhouse Effect. The most significant for many organisations is the emission of greenhouse gases from power stations supplying them with electricity. But in addition, other sources include things like staff transport to work, the emissions generated during waste disposal operations dealing with waste generated at site (transport and waste management related plus methane from the breakdown of any organic materials landfilled or incinerated).

Indirect supply chain emissions

The manufacture of any materials used within the manufacturing process as well as in the construction of the buildings in which the company operates will have a whole chain of energy, transport and waste related emissions comparable to those outline above. In addition, if primary raw materials come from sources where there may have been clearance of forest areas (particularly in tropical areas) then there is the added impact of removal of carbon sinks which will further contribute to increasing concentrations of greenhouse gases in the atmosphere.

Direct product based contributions

The manufacturing of products inevitably results in activities by the end user that may contribute to greenhouse gas emissions via energy use, repairs or maintenance, usage related transport or simply end of life disposal of packaging and the product itself.

Full answer – part b)

Life cycle analysis considers the environmental burdens (including greenhouse gas emissions) from all stages of the life cycle of a product from raw material extractions through component supply chains to the main manufacturing process, product usage and final disposal.

In the inventory stage of the LCA process key environmental burdens are quantified at each stage of the life cycle with reference to the common unit of the final product. In the impact assessment stage this data allows a summing up of the total greenhouse gases generated at each life cycle stage and also across the life cycle as a whole. When expressed across the whole life cycle the resulting quantity of all greenhouse gases emitted directly or indirectly (normalised to carbon dioxide equivalent) may be stated as a single number often referred to as the 'embedded carbon' of the product (if focused upstream of the end user) or the 'carbon footprint' of the product (if using the data from the whole life cycle including in use and disposal data).

Clearly there is scope for error in such calculations due primarily to the availability of reliable information and allocation procedures for 'sharing out' greenhouse gas emissions in the supply chain. However, the LCA technique may add greatly to the understanding of where in the life cycle the greatest burdens lie in terms of greenhouse effect contributions and particularly in the testing of the impact reduction potential of proposed improvements such as supplier changes or product/packaging redesign.

Chapter 9 Question 3

A chilled foods manufacturer is in the process of inviting tenders for a contract for the servicing and maintenance of refrigeration systems at its site. Amongst other requirements, the contract will include the removal of all related wastes including electrical equipment, refrigerant gases and oils.

Describe *the* environmental *issues the organisation should consider* when selecting a contractor. *[20]*

Answer plan

Describe – detailed factual account of key features

General selection criteria:

- Environmental policy/ISO14001
- Compliance history
- Improvement programmes/environmental performance reporting
- Location of contractor organisation (local vs distant)
- Trade experience/references

Specific selection criteria:

- Staff competence under f-gas regs
- Waste carriers license and disposal arrangements
- Containment/pollution prevention measures in method statements

Full answer

There are a number of issues that the organisation should consider at the tendering stage for this type of contract. They may be considered as general selection criteria and contract specific criteria.

General selection criteria

For all contracts the following issues should be considered to increase the likelihood of working with a reliable contractor that will achieve the aims of the contract with minimum environmental risk and with potential for impact reduction. Selecting contractors that have a robust and credible environmental policy that is relevant to their organisation and scope of work is a basic minimum. If possible, and appropriate to the organisation, evidence of an effective environmental management system being in place would be desirable, ideally certified by a third party to a recognised standard, e.g. 14001.

Contractors should be quizzed on their compliance history and in particular any outstanding prosecutions or enforcement notices that might be in place. Contractors with a clean history would clearly be preferred.

Credit should also be given to contractors that can not only demonstrate a good compliance record, but who can also show evidence of actively reducing their environmental impact, for example through waste reduction or energy efficiency programmes. If contractors can demonstrate that they are offering a service that has environmental benefits over competitors they should be given additional credit.

The location of the contractor's base may also be a selection criterion. Local suppliers might be expected to have lower transport related impacts than those situated more remotely. Although this ought to be considered in relation to the specifics of the contract involved as it may be of lesser or greater importance depending on the nature of the contract and the relative significance of contract related transport.

Finally evidence of trade experience and customer references should be sought as part of the tendering process.

Contract specific selection criteria

Under the Fluorinated Greenhouse Gas Regulations, 2009, personnel working with listed substances, which include many of the gases currently used in air conditioning and refrigeration systems, are required to be able to demonstrate technical competence. For this contract, tendering companies should be expected to demonstrate a familiarity with the Regulations and the associated Defra guidance and be able to name 'competent personnel' that will be involved in carrying out maintenance and servicing activities on site.

In addition, due to the waste disposal element of the contract, contractors should be able to provide waste carrier registrations issued by the Environment Agency under the Waste (England & Wales) Regulations 2011. As part of this area of competency, evidence of their knowledge and arrangements for the hazardous classification of both waste oils and refrigerant gases should be reviewed, as well as the planned arrangements for final disposal or recycling to ensure that all necessary parties are in possession of an appropriate Environmental Permit under the Environmental Permitting Regulations 2010.

Finally, contract specific method statements that clearly delineate responsibilities and arrangements for pollution prevention and incident response should be reviewed for credibility and clarity.

In many tendering processes a scoring scheme or decision tree might prove useful in comparing information provided by a number of contractors against the range of criteria described above.

Chapter 10 Question 1

a) ***Outline*** *the key elements of the so-called 'waste hierarchy'. [8]*

b) *Using suitable examples,* ***describe*** *how the 'waste hierarchy' could be implemented in relation to office waste. [12]*

Marking scheme	Available marks
Part a)	
Linked to UK waste strategy and based on perceived level of environmental impacts	1
For each of the following define and explain impact/benefit in terms of disposal and resource consumption:	
Reduce – eliminate	2
Reuse	2
Recycle/recover/compost	2
Dispose	2
Regulatory context provided by the Waste (England and Wales) Regulations 2011	1
Part b)	
Examples should be across the following categories with marks limited for repeated examples within the same category:	

Marking scheme	Available marks
Reduce – e.g. construction design to minimise waste, avoid over ordering, ensure appropriate materials storage to avoid yard waste, bespoke rather than standard sizing of materials	Up to 3
Reuse – e.g. cut and fill techniques with soil, returnable packaging, return of unused construction materials to suppliers	Up to 3
Recycle – e.g. aggregates recycling or production form brick and concrete (on or off project), segregation of metals, cardboard, wood etc. for off project recycling, waste oil	Up to 3
Recover – e.g. burning of waste wood in biomass boilers, sending general waste to municipal waste to energy plants	Up to 2
Compost – e.g. canteen or landscaping green waste	Up to 2
Dispose – landfill of remaining wastes	1

Chapter 11 Question 1

***Identify** four reasons why an organisation may choose to make public an annual report on their environmental performance and outline the typical content that might be included within such a report. [20]*

Full answer

Drivers to publish environmental reports include:

- Government/statutory requirements – several countries have some kind of mandatory corporate environmental reporting. The UK for example requires all publically listed companies to report annually on their greenhouse gas emissions in a format that meets DEFRA guidance.
- Shareholder, customer pressure – these pressures are often manifested through listings on review schemes such as the Business in the Community sustainability index or the Dow Jones sustainability index. Rankings in such schemes are often considerably enhance by public performance reporting.
- Accreditation to the EU eco-management and audit scheme (EMAS) which requires public reporting as well as a management system aligned to the requirements of ISO14001.
- Reputation management – organisations subject to considerable media and public scrutiny such as oil companies and

power generators may be inclined to publically report their environmental performance in order to present a balanced view of their efforts and achievements.

There is considerable guidance now available on the recommended content of a corporate environmental report from organisations such as the Global Reporting Initiative, World Business Council for Sustainable Development, DEFRA and of course the EU eco management and audit scheme (EMAS). While there is some variation in the scope and wording used, the following areas might be consider a typical minimum for a good standard report:

- A description of the company activities and the context in which it operates including locally sensitive receptors if applicable, and changes in company activities since any previous report.
- A description of key environmental aspects and impacts with a clear statement of the scope of influence to which the report relates, e.g. direct impacts only or supply chain and product related impacts too. This scope setting is particularly important in relation to statements around total company impact, e.g. in relation to greenhouse gas emissions. The guidance allows for variable scope but demands transparency to avoid misleading statements.
- A description of key policies, objectives and targets related to the environmental impacts of the organisation.
- Performance data in the form of key performance indicators relating to significant environmental aspects. These may take the form of emissions/waste data in absolute or normalised format, number of incidents, complaints etc. or in a more positive format, e.g. reductions in emissions or impacts associated with successful actions completed during the reporting period. Alternatively, they may be in the form of management performance indicators such as the degree of compliance demonstrated with internal control standards. In all cases the guidance suggests that performance data should be reported in a balanced and accurate manner and where possible be subject to third party assurance auditing. Good practice also suggests that performance data should also be presented in an easy to understand format that is a fair representation of the status of the organisation.

- Details of any assurance auditor including credentials and independence. The GRI guidelines allow for a variable statement of compliance depending on the degree of assurance process applied by the organisation. Again the idea is that there is transparency in relation to the accuracy and reliability of the data included in the report to the reader.

Chapter 12 Question 2

Many organisations include environmental awareness in their company induction training for new employees.

a) ***Outline*** *the environmental content that should be included in such a programme. [10]*

b) ***Describe****, with appropriate examples, ways in which all employees might be encouraged to engage with a company's environmental management programme, with particular reference to procedure compliance. [10]*

Answer plan

a) Outline – give the most important features of:

- policy and procedures
- legal requirements
- key aspects and company drivers (business benefits)
- reputation standards, e.g. ISO14001
- requirements of employees
- consultation and disciplinary processes
- statement about the limitation of one off induction briefings

b) Describe – give a clear picture of:

- management commitment and regular visible discussion
- performance feedback
- audits and inspections
- involvement in preparation and updating of procedures and local aspect evaluations
- non-conformance reporting and individual disciplinary measures
- regular refresher briefings/awareness events

Full answer

a) Induction programmes are often used to provide all employees with a basic overview of company environmental policy and key issues relevant to them. Content that might be relevant to a manufacturing organisation might include the following:

- Company policies and commitments and perhaps as importantly where to get further information about issues that may be encountered.
- Legal and other commitments – often associated with policy this relates to specific requirements placed on the business by regulators in the form of consents or permits, customers in the form of contractual requirements and parent companies in the form of policies, objectives and targets. The aim in an induction programme would not be to go through all of these requirements in detail, merely to give an overview of the pressures on the organisation to manage its environmental issues effectively. It may also be possible to highlight some of the potential benefits of the programme from the business perspective, e.g. savings in raw materials, waste and energy costs, market advantage etc.
- Key environmental aspects, impacts and receptors – where possible with specific emphasis on those aspects relevant to the job and work location of the new employees. This might be done in two parts with an initial overview of company priorities at a general company induction and then a follow up specific briefing during a secondary work place induction.
- Management systems and assurance processes – highlighting any formal systems such as ISO14001 and the process of routine accreditation checks by independent auditors. Within this, highlighting the use of the system by the company to record key issues, controls and procedures. As with the aspects briefing this might be best done as a two part process – generic overview and then job specific briefings as a follow up. The latter could include a basic synopsis of these are the key environmental issues directly relevant to you and these are the procedures you need to follow.
- Responsibilities of all employees – this entails clarity over what is expected of them – both in generic terms such as reporting

incidents or uncontrolled risks, and specifically in following work place procedures. This may also be a place to highlight sources of support or consultation in the event of uncertainty around an environmental issue.

- Finally some statement about disciplinary proceedings and top management commitment may help to provide weight to the message delivered.

b) There are many ways that employees could be encouraged to follow company procedures, including the following:

- Management commitment and regular visible discussion of environmental matters can help ensure that the environmental controls are seen as important to those to whom employees directly report.
- Performance feedback in the form of incident and audit reports, name and shame lists and/or raise and praise lists to highlight poor or good performance.
- Audits and inspections – carried out by local staff as well as the company specialists (environmental auditors etc.). Ideally audits become less a policing exercise and more an assurance opportunity identification process, with employees engaging with auditors to share concerns or hear about good practice elsewhere.
- Involvement in preparation and updating of procedures and local aspect evaluations – this gives a sense of ownership and understanding of what is required and why. It reduces the likelihood of resistance to compliance with procedures on the grounds that they are inappropriate or in some other way difficult to work with.
- Non-conformance reporting and individual disciplinary measures – where necessary individuals and sometimes whole departments need to be able to see that while efforts are made to make procedures workable and reasonable, there are consequences if they are not complied with.
- Regular refresher briefings/awareness events – to ensure that practices and the reasons for them does not get forgotten over time.

Inductions are probably the most common way to communicate basic company information including that relating to environmental issues. They are also one of the least reliable ways to ensure that employees know about issues relevant to them and are clear about what their role is in managing them. As indicated above keeping an initial induction briefing to the bare minimum maximises the chance of the key information being retained. Subsequent follow up on a more job specific level can then be used to add detail.

APPENDIX 3

Glossary of terms

The following terms are used in the text and are commonly encountered in the environmental management context.

Abatement	Control or reduction of pollution or removal of a nuisance by engineering (plant, equipment), operational (procedure) or regulatory (licence or order) means.
Acid rain	The reaction of acid gases (especially sulphur dioxide and oxides of nitrogen) with atmospheric moisture thereby increasing the acidity of resulting precipitation. Such acidic precipitation over time results in increased acidity of receiving waters and soil, particularly in areas with naturally acidic soils and/or vegetation.
Ambient pollution	Background levels of pollution that reflect the cumulative result of pollution from all relevant sources.
Aquifer	Underground geological formation containing water that has typically accumulated over many years. Often used for potable water supply.
Best available techniques (BAT)	Performance standard required under an Environmental Permit. 'Best' refers to the most effective techniques in achieving a high overall level of environmental protection. 'Available' means proven techniques allowing implementation in the relevant sector under economically and technically viable conditions. 'Techniques' refers both to technology used and the way in which the installation is designed, built, maintained, operated and decommissioned.

Best practicable environmental option	The option which provides the most benefit or least damage to the environment as a whole, at acceptable cost, in the long term as well as the short term. The term recognises that in abating environmental impacts there are often trade-offs (e.g. reducing air pollution may increase production of solid waste or effluent) which may vary significantly depending on local receptors.
Bio-accumulation	The build-up of persistent pollutants in the body tissues of some organisms, frequently increasing in concentration through the food chain.
Biochemical oxygen demand (BOD)	A measure of the pollution in a body of water or effluent stream, based on the organic material it contains. Such organic material provides food for aerobic bacteria which require oxygen to break down their food source. The greater the volume of organic material, the higher the concentration of bacteria and the greater the oxygen demand. (BOD values give no indication of other pollutants such as suspended sediments or heavy metals). If BOD exceeds available dissolved oxygen, oxygen depletion occurs and many aquatic organisms suffer – fish kills are not uncommon in such circumstances.
Biodiversity	The variety of life on earth, considered at the level of habitats, plant and animal species and in terms of genetic diversity within species.
Biomass and biomass energy	Biomass is the total weight of living organic matter in a given area. Biomass energy is the energy available from organic material including wood, crops, crop residues, industrial and municipal organic waste, food processing waste and animal wastes. They may be used directly or converted before use (e.g. conversion of sugar beet waste into alcohol). Biomass energy is renewable if managed appropriately but as with fossil fuels produces smoke, soot and carbon dioxide when burnt.
Bund	A type of 'secondary containment' in the form of an impervious wall around a tank or other primary container, designed to prevent pollution in the event of a leak or spill from the primary container.
Carbon footprint	Defined by the Carbon Trust as 'the total set of greenhouse gas (GHG) emissions caused by an organisation, event, product or person'.

Chemical oxygen demand (COD)	A measure of the oxygen consumed in the chemical oxidation of organic and inorganic matter in water or effluent. It provides an indication of the impact of effluent on dissolved oxygen levels. A standard test uses potassium dichromate as an oxidising agent and measures oxygen consumed over a standard period of time.
Combined heat and power (CHP) (also known as cogeneration)	Energy technology which utilises the heat generated during electricity generation for space/water heating or steam generation. CHP may achieve energy generation efficiencies in excess of 85% compared with typical coal fired power generation efficiencies of 30–40% (i.e. 30–40% of the energy available in the fuel is converted into usable energy).
Corporate social responsibility (CSR)	A form of corporate self-regulation integrated into a business model whereby business monitors and ensures its active compliance with the spirit of the law, ethical standards and international norms. The goal of CSR is to embrace responsibility for the company's actions and encourage a positive impact through its activities on the environment, consumers, employees, communities and other stakeholders. CSR is often associated with 'triple bottom line' reporting whereby the business gauges its success in terms of positive results in relation to people, planet and profit.
Cumulative impacts	A combination of impacts arising from either: similar environmental aspects from different sources occurring simultaneously, or sequential aspects from the same source or different sources that do not allow adequate time for environmental recovery between occurrences. In either case the sum of the impacts is greater than each individual environmental change.
Data	Information that might be used to assess priorities and monitor performance. The following key terms are often used: **Qualitative data** – information based on opinion, judgement or interpretation. **Quantitative data** – numerical measurements that may be subject to statistical analysis or benchmarking. **Absolute data** – the actual amount of something measured, e.g. tonnes of waste generated per year. **Normalised data** – based on absolute data but subject to some kind of referencing process to enable comparison between locations or over time.

Data verification and assurance	Assurance is defined as a formal guarantee – a positive declaration that a thing is true. Verification may be defined as the process of establishing the truth or accuracy of data. So verification processes provide assurance. Other relevant terms from accounting that are applied universally to any kind of data verification activities include: **Materiality** – a concept from accounting that relates to the importance, significance and relevance of information including the degree of accuracy of the data. **Responsiveness** – demonstrating reactivity to stakeholder concerns and meeting their information needs. **Completeness** – covering and including all relevant sources and activities relevant to an organisation in terms of environmental performance.
Eco-efficiency	Defined by the World Business Council for Sustainable Development as 'the delivery of competitively priced goods and services that satisfy human needs and bring quality of life, while progressively reducing ecological impacts and resource intensity throughout the life cycle to a level at least in line with the earth's estimated carrying capacity'.
Eco-Management and Audit Scheme (EMAS)	The environmental management system model produced by the European Union as a voluntary standard that may be subject to third party verification. Since its revision in 2001 it has become closely aligned with ISO 14001.
Ecosystem	A community of interdependent organisms and the physical and chemical environment that they inhabit.
Ecosystem services	A term used to describe the value to humans of various aspects of natural systems. Normally expressed in relation to supporting, provisioning, regulating and cultural services.
Eco-toxicity	The potential for biological, chemical or physical stressors to affect ecosystems.
Ecological footprint	A measure of human demand on the Earth's ecosystems. It represents the amount of biologically productive land and sea area necessary to supply the resources a human population consumes and to mitigate associated waste.
Effluent	Liquid waste stream typically released to drain or water body.

Emission	Term applied to the release of any waste substance or noise but most frequently used to describe releases to atmosphere.
Environment	Surroundings in which an organisation operates, including air, water, land, natural resources, flora, fauna, humans and their interrelation. These can extend from within the organisation to the global system.
Environmental aspect	Those elements of an organisation's activities, products and services which can interact with the environment.
Environmental audit	Term used in a number of contexts to describe a variety of investigations into the environmental performance of a site or organisation. The scope of investigation and level of detail varies widely.
Environmental impact	Any change in the environment, whether adverse or beneficial, resulting from an organisation's activities, products or services.
Environmental impact assessment	The formal assessment and analysis of the potential impact of various forms of human activity on the environment. Specialist studies of existing environmental conditions, expected changes and proposed mitigation measures are legally required as part of the planning application process for defined development projects.
Environmental management system	Those elements of an organisation's overall management system that monitor, control and improve performance against an organisation's policy or priorities. Internationally recognised models of management systems include ISO 14001 and EMAS.
Environmental quality standard	A standard typically set by a regulatory body that specifies the quality of the ambient environment, e.g. the maximum concentration of a pollutant in the air or a body of water.
Eutrophication	The process of algae proliferation in water due to a high concentration of nutrients. Often associated with run-off or effluent containing high concentrations of fertilisers or rapidly decomposable organic matter.
Fugitive emission	Ad hoc releases of emissions (normally to air), e.g. from pipe/joint leaks, evaporative losses from storage tanks or during uncontained usage. Particularly relevant for volatile organic compounds.

Greenhouse effect	The term used to describe the selective response of the atmosphere to different types of radiation. Incoming short wave radiation passes through unaltered, while returning long wave radiation is absorbed by the 'so called' greenhouse gases.
Greenhouse gas (GHG)	The group of about 20 gases responsible for the greenhouse effect through their ability to absorb long wave radiation. Carbon dioxide (CO_2) is the most abundant but methane (CH_4), nitrous oxide (N_2O), the chlorofluorocarbons (CFCs) and tropospheric ozone also make significant contributions to the greenhouse effect.
Groundwater	Water that accumulates in the pore spaces and rocks below the earth's surface. It originates as precipitation and percolates down into aquifers or finds its way back to the surface via springs or connection to surface water courses. The upper limit of groundwater saturation is known as the water table.
Habitat	The specific environment in which an organism lives. Although the term is often used in relation to a particular species, any given environment will be shared by a variety of interdependent organisms.
Hydrocarbon	Organic compounds composed of hydrogen and carbon. They may be solid (e.g. coal), liquid (e.g. crude oil) or gaseous (e.g. natural gas) in form. They are used primarily as fuels but are also utilised as lubricants and as feed stocks for a variety of industrial materials (especially important to the plastics and fertiliser industries).
Indirect environmental impact	Any change in the environment, whether adverse or beneficial, that is caused by third parties acting on behalf of, or in support of, an organisation's activities, products or services, e.g. power generation impacts associated with the supply of energy to an organisation for use in its manufacturing process.
ISO 14001	The environmental management system model produced by the International Standards Organisation against which organisations may choose to be assessed by third party certification bodies.
Key performance indicator (KPI)	A parameter that measures the level of achievement in an area considered to be of particular importance to an organisation. Ideally KPIs will be normalised, i.e. referenced to some activity indicator to allow comparison over time, e.g. kilogrammes of waste produced per product manufactured.

Life cycle analysis	The compilation and evaluation of the inputs, outputs and the potential environmental impacts of a product system throughout its life cycle, i.e. through the consecutive and interlinked stages of a product system, from raw material acquisition or the generation of natural resources to the final disposal.
Nuisance	Interference with a person's use and enjoyment of the environment through something that annoys, bothers or causes damage to that person or their property. Includes noise, odour and visual intrusion.
Oxides of nitrogen	Term used to describe a group of gases (nitrous oxide N_2O, nitric oxide NO and nitrogen dioxide NO_2) formed by the combination of nitrogen and oxygen under high energy conditions (typically associated with high temperature incineration and internal combustion engines).
Ozone depletion	The breakdown of ozone in the upper atmosphere (stratosphere) by man-made chemicals containing bromine and chlorine, e.g. CFCs. This layer of ozone acts as a protective layer to life on earth, filtering out harmful ultraviolet radiation from the sun.
PANs	Peroxyacetyl nitrates, or PANs, are powerful respiratory and eye irritants present in photochemical smog. PANs are secondary pollutants, which means they are not directly emitted as exhaust from power plants or internal combustion engines, but they are formed from other pollutants by chemical reactions in the atmosphere.
Photochemical smog	Smog produced by chemical reactions in the presence of sunlight on pollutants arising from combustion especially hydrocarbons and oxides of nitrogen (from vehicles). A variety of toxic chemicals are produced in the 'smog' including ozone (harmful to plants; irritant to eyes, nose and throat), aldehydes (smelly, poisonous and irritant to eyes, nose and throat) and peroxyacetyl nitrate (PAN – toxic; irritant to eyes, nose and throat).
Photo-synthesis	The biochemical process by which plants use energy from sunlight to convert carbon dioxide and hydrogen (from water) into simple carbohydrates.
Policy	An agreed approach to a particular issue or group of issues that may be enacted or achieved through a number of policy instruments (see below). Policies may be set at international, national, corporate and even individual scales.

Policy instrument	Mechanisms used to achieve or implement a particular policy. Typical examples include fiscal (e.g. taxation), legislative (e.g. a new law), market (e.g. product design specifications) or voluntary instruments (e.g. ISO14001).
Pollution	The introduction by man of substances or energy into the environment that are liable to cause hazards to human health, harm to ecological systems, damage to structures or amenity, or interference with legitimate uses of the environment.
Pollution linkages	The knock on effects of pollution through physical or biological processes. Also described under the concept of pollution pathways, i.e. an initial source, pathway, receptor linkage then causing secondary impacts in other receptor groups.
POPs – persistent organic pollutants	A group of chemicals that remain unchanged in the environment for many years and which may find their way into the food chain and human tissues via bioaccumulation. The so called 'dirty dozen' POPs include the pesticides aldrin, chlordane, DDT, dieldrin, endrin, heptachlor, mirex and toxaphene, the industrial chemicals polychlorinated biphenyls (PCBs), hexachlorobenzene and the combustion by-products dioxins and furans.
Preliminary environmental review	The process of evaluation of legal requirements, management standards and environmental priorities for an organisation. Often carried out at the beginning of an environmental management system implementation programme and involving a 'gap analysis' with a relevant benchmark (typically ISO 14001).
Receptor	An entity that can be subject to an environmental impact caused by pollution or resource consumption. Includes water bodies, air, land, communities, ecosystems, individual organisms, human beings and property. Some receptors are particularly sensitive to certain impacts and are known as 'sensitive receptors'.
Risk	The potential for realisation of unwanted, adverse consequences to human life, health, property or the environment (a combination of the likelihood and consequences of a specific outcome, good or bad).
Secondary environmental impact	Any change in the environment arising as a direct consequence of an initial change caused by an organisation's activities, e.g. fish kill resulting from water pollution caused by a discharge by an organisation.

Sensitive receptors	Receptors that, for some reason, are particularly susceptible to impacts resulting from an organisation's activities. They may be human, e.g. children or the elderly, flora or fauna, e.g. rare species or habitats, or physical, e.g. aquifers used for potable water supply.
Significance assessment	A systematic evaluation of the relative importance of an organisation's environmental aspects and impacts. It includes the assessment of each aspect in relation to a standard set of criteria related to regulatory requirements and stakeholder concerns. The process results in a logical and repeatable prioritisation of aspects which facilitates management decisions on controls and improvements.
Stakeholder	Individuals, communities or organisations that have an interest in an organisation or who are affected by its policies, practices and performance.
Storm water	Storm water is water that originates during precipitation events. It may also be used to apply to water that originates with snowmelt that enters the storm water system. Storm water that does not soak into the ground becomes surface runoff, which either flows directly into surface waterways or is channelled into storm sewers, which eventually discharge to surface waters.
Stratosphere	The stratosphere is the second major layer of Earth's atmosphere, just above the troposphere, and below the mesosphere. The stratosphere is situated between about 10km (6mi) and 50km (30mi) altitude above the surface at moderate latitudes, while at the poles it starts at about 8km (5mi) altitude.
Sustainability	Sustainability is a description of environmental balance where resource consumption and the absorption of wastes are balanced by its natural rate of replenishment and breakdown – the term 'environmental sustainability' is sometimes used to emphasise this meaning.
Sustainable development	Sustainable development is the process of transforming human society into a form that allows 'sustainability' to exist. It involves consideration of the social and economic issues (as well as the environmental ones) that need to be addressed along the way.
Trade effluent	Relates to effluent generated from a commercial or industrial process. Pollutant loads and discharge routes may vary widely. Typically, however, these discharges are subject to stricter legal controls than storm water runoff.

Troposphere	The troposphere is the lowest portion of Earth's atmosphere. It contains approximately 75% of the atmosphere's mass and 99% of its water vapour. The average depth of the troposphere is approximately 17km (11mi) in the middle latitudes. It is deeper in the tropical regions, up to 20km (12mi), and shallower near the poles.
Value chain	A term coined by Michael Porter to describe the activities involved in transforming raw materials into usable and saleable products including all raw material extraction/ production processes, manufacturing and transport processes plus selling and distribution activities. These are the company operations that mirror the 'life cycle' of a product with each step 'adding value'.
Volatile organic compounds (VOCs)	Organic compounds which evaporate readily including acetone, ethylene, benzene, propylene and many solvent preparations. VOCs contribute to a number of air quality issues either directly (e.g. benzene is carcinogenic) or indirectly as components of photochemical smog.

APPENDIX 4

Legislation summary and revision tool

The purpose of the following sheets is to assist with understanding as well as revision and exam preparation in relation to the legal issues part of the IEMA and NEBOSH syllabi. No 'completed version' is provided as the process of completing the sheet with key phrases meaningful to you is a large part of the process. Please note that not every piece of legislation described in the accompanying *Manual of Environmental Management* is listed here. This is intentional, this listing is focused on only the most widely applicable legislation.

Legislation	**Summary of requirements**	**Responsible body**	**Offence/ legal breach**
Anti-Pollution Works Regulations 1999			
Clean Air Act 1993			
Climate change Levy regulations, 2001			

Appendix 4 Legal summary table

Legislation	Summary of requirements	Responsible body	Offence/ legal breach
Contaminated Land Regulations 2006 [also Scotland (2000) and Wales (2006)]			
Control of Major Accident Hazards 1999 (as amended)			
Control of Pesticides Regulations 1986 (as amended)			
Control of Pollution (Oil Storage) Regulations, 2001			
Control of Pollution Act (amendment) 1989 – section 60/61 (construction noise)			
End of Life Vehicle Regulations, 2003			
Environment Act 1995 Part IV – air quality			
Environment Act 1995 Part V – producer responsibility framework			
Environmental Assessment of Plans and Programmes Regulations 2004			
Environmental Damage (Prevention and Remediation) Regulations 2009			
Environmental Permitting Regulations 2010			
Environmental Protection (Disposal of Polychlorinated Biphenyls and other Dangerous Substances (England and Wales) Regulations 2000			

Legislation	Summary of requirements	Responsible body	Offence/ legal breach
Environmental Protection Act 1990 Part II section 34 – the Duty of Care			
Environmental Protection Act 1990 Part III – statutory nuisance			
Environmental Protection (Controls on Ozone-Depleting Substances) Regulations 2011			
Fluorinated Greenhouse Gases Regulations 2008			
Greenhouse Gas Emissions Trading Scheme Regulations, 2003			
Habitats Directive and Conservation (Natural Habitats etc.) Regulations 1994			
Hazardous Waste Regulations 2005			
Noise and Statutory Nuisance Act 1993			
Noise Emission in the Environment by Equipment for Use Outdoors Regulations, 2001			
Planning (Hazardous Substances) Act 1990 (as amended)			
Producer Responsibility Obligations (Packaging Waste) Regulations 2010			
Protection of Badgers Act 1992			
REACH Enforcement Regulations 2008			

Legislation	Summary of requirements	Responsible body	Offence/ legal breach
Restriction of the Use of Certain Substances in Electrical and Electronic Equipment Regulations, 2006.			
Town and Country Planning Act (Environmental Impact Assessment) (England & Wales) Regulations 1999			
Town and Country Planning Act 1990 (as amended by the Planning and Compensation Act 1991)			
Trade Effluent (Prescribed Processes and Substances) Regulations 1989 (as amended)			
Waste Batteries and Accumulators Regulations 2009			
Waste Electrical and Electronic Equipment Regulations 2006			
Waste (England & Wales) Regulations 2011			
Water Industry Act 1991			
Water Resources Act 1991 (WRA 1991)			
Wildlife and Countryside Act, 1981 (as amended by the Countryside and Rights of Way Act 2000)			

Bibliography

Accountability (2008) AA1000 Assurance Standard, Accountability.

Beaman and Kingsbury (1981) 'Assessment of nuisance from deposited particulates using a simple and inexpensive measuring system', *Clean Air*, 11(2): 77–81.

Brady, J., Ebbage, A. and Lunn, R. (2011) *Environmental Management in Organizations: The IEMA Handbook*, 2nd edn. Earthscan. Available at: http://www.iema.net/suggested-reading#sthash.rqUp6uxx.dpuf.

British Standards Institute (1997) BS 4142:1997 Method for rating industrial noise affecting mixed residential and industrial areas.

British Standards Institute (2003) BS 7445-1:2003 Description and measurement of environmental noise: Guide to quantities and procedures.

British Standards Institute (2010) BS 8903:2010 Principles and framework for procuring sustainably.

Dale, R.A. (trans.) (2002) *Tao Te Ching*. London: Watkins.

DEFRA (2010) *Odour Guidance for Local Authorities*. London: DEFRA.

DEFRA (2011) *Green Claims Guidance*. London: DEFRA.

DEFRA (2012) *Environmental Protection Act 1990: Part 2A. Contaminated Land Statutory Guidance*. London: DEFRA.

DEFRA (2012) *Biodiversity Offsetting: Guidance for Developers*. London: DEFRA.

DEFRA (2013a) *Encouraging Businesses to Manage Their Impact on the Environment*. London: DEFRA.

DEFRA (2013b) *Environmental Reporting Guidelines – Including Mandatory Greenhouse Gas Reporting Guidance*. London: DEFRA.

DEFRA (2013c) *Making Sustainable Development a Part of All Government Policy and Operations.* London: DEFRA.

DEFRA (2014) *Reducing and Managing Waste.* London: DEFRA.

EMAS (2009) EC Eco-management and audit scheme.

Environment Agency (2002) *Horizontal Guidance note H3 Part 2 Noise Assessment* and *control.* Bristol: Environment Agency.

Environment Agency (2011) *Above Ground Oil Storage Tanks: Pollution Prevention Guidelines PPG2.* Bristol: Environment Agency.

Environment Agency (2013) *Technical Guidance WM2.- Hazardous Waste: Interpretation of the Definition and Classification of Hazardous Waste*, 3rd edn. Bristol: Environment Agency.

Environment Agency (2013) Environmental Permitting Regulations Operational Risk Appraisal (OPRA for EPR) version 3.8.

Environment Agency (2011) *Horizontal Guidance Note H4 Odour Management: How to Comply with Your Environmental Permit.* Bristol: London: Environment Agency.

Environmental Protection UK (2010) *Pollution Control Handbook: The Essential Guide to UK & European Pollution Control Legislation.*

Global Reporting Initiative (2013) G4 – Sustainability Reporting Guidelines. Reporting Principles and Standard Disclosures. GRI.

Hanson, C. et al. (2012) *The Corporate Ecosystems Review: Guidelines for Identifying Business Risks and Opportunities Arising from Ecosystem Change.* Washington, DC: World Resources Institute.

Henriques, A. (2001) *Sustainability: A Manager's Guide.* London: British Standards Institute.

Hyde, P. and Reeve, P. (2005) *Essentials of Environmental Management.* Wigston: Institute of Occupational Safety and Health.

IEMA (2006) *Risk Management for the Environmental Practitioner.* Lincoln: IEMA.

ISO14001:2004 Environmental management systems – requirements with guidance for use.

ISO14004:2004 Environmental management systems – general guidelines on principles, systems and support techniques.

ISO14005:2010 Environmental management systems – guidelines for the phased implementation of an environmental management system, including the use of environmental performance evaluation.

ISO14006:2011 Environmental management systems – guidelines for incorporating ecodesign.

ISO14031:2013 Environmental management – Environmental performance evaluation – guidelines

ISO14040:2006 Environmental management – life cycle assessment – principles and framework

ISO 19011:2011 – Guidelines for auditing management systems.

Kemp, D. (1998) *The Environment Dictionary*. London: Routledge.

Owen, A. (2000) *Ecodesign: A Training Guide for Business*. Sheffield: Green Training Works Ltd.

Porritt, J. (2007) *Capitalism as if the World Matters*. London: Routledge.

Reeve, R. (2002) *An Introduction to Environmental Analysis*. Chichester: John Wiley & Sons Ltd.

United Nations Conference on Environment & Development Rio de Janeiro, Brazil, 3 to 14 June 1992 – AGENDA 21.

UNEP (2012) *UNEP Global Outlook on Sustainable Production and Consumption Policies*. New York: UN.

World Business Council for Sustainable Development (2011) *Guide to Corporate Ecosystem Valuation*. Geneva: World Business Council for Sustainable Development.

Walker, P. (2006) *Change Management for Sustainable Development*. Lincoln: IEMA.

Waters, B. (2013) *Introduction to Environmental Management*. London: Routledge.

Welsh Government (2013) *One Wales: One Planet: The Sustainable Development Annual Report 2012–13*. Cardiff: Welsh Government.

Index

References in **bold** indicate a table, those in *italics* are for figures, and glossary terms are shown in **bold**, **underlined**.